Wavelets:
A Mathematical
Tool for Signal
Analysis

SIAM Monographs on Mathematical Modeling and Computation

About the Series

In 1997, SIAM began a new series on mathematical modeling and computation. Books in the series develop a focused topic from its genesis to the current state of the art; these books

- present modern mathematical developments with direct applications in science and engineering;
- describe mathematical issues arising in modern applications;
- develop mathematical models of topical physical, chemical, or biological systems;
- present new and efficient computational tools and techniques that have direct applications in science and engineering; and
- illustrate the continuing, integrated roles of mathematical, scientific, and computational investigation.

Although sophisticated ideas are presented, the writing style is popular rather than formal. Texts are intended to be read by audiences with little more than a bachelor's degree in mathematics or engineering. Thus, they are suitable for use in graduate mathematics, science, and engineering courses.

By design, the material is multidisciplinary. As such, we hope to foster cooperation and collaboration between mathematicians, computer scientists, engineers, and scientists. This is a difficult task because different terminology is used for the same concept in different disciplines. Nevertheless, we believe we have been successful and hope that you enjoy the texts in the series.

Joseph E. Flaherty

Wavelets: A Mathematical Tool for Signal Processing

Charles K. Chui

Texas A&M University
College Station, Texas

Society for Industrial and Applied Mathematics

Philadelphia

Library of Congress Cataloging-in-Publication Data

Chui, C. K.
   Wavelets : a mathematical tool for signal analysis / Charles K.
Chui.
   p. cm. -- (SIAM monographs on mathematical modeling and
computation)
   Includes bibliographical references (p. -) and index.
   ISBN 0-89871-384-6 (pbk.)
   1. Wavelets (Mathematics) 2. Signal processing--Mathematics.
I. Title. II. Series.
QA403.3.C49 1997
621.382'23'015152433--dc21   96-51635

siam is a registered trademark.

Contents

Foreword

There are two key operations on a wavelet $w(t)$. One is a shift, to $w(t - 1)$ and $w(t - 2)$ and $w(t - k)$. The other is a rescaling, to $w(2t)$ and $w(4t)$ and $w(2^j t)$. A combination of shifting and scaling produces a whole family of wavelets $w(2^j t - k)$, all coming from the single function $w(t)$. Can we choose this function so that all these shifted and rescaled functions are *orthogonal* to each other? Can we choose two wavelets $w(t)$ and $\widetilde{w}(t)$ so that the two families are *biorthogonal*?

$$\text{Orthogonal}: \int w_{jk} w_{JK}\, dt = 0 \quad \text{unless } j = J \text{ and } k = K.$$

$$\text{Biorthogonal}: \int w_{jk} \widetilde{w}_{JK}\, dt = 0 \quad \text{unless } j = J \text{ and } k = K.$$

Ten years ago, this hardly seemed possible. Only the simplest constructions were understood. To be useful in practice, the wavelets have to produce a "good basis." Other signals and other functions must be well represented by the wavelets. If we can come close to a typical signal with just 5% of the wavelets, we have compression of 20:1. And if a concentration of energy near a specific j, k can locate a signal (in time as well as frequency!), we have an efficient detector.

These goals are now achievable. The constructions begin with a highpass filter H_1 and a lowpass filter H_0. Wavelets are built from filters! The choice of a finite (small) set of coefficients, $h_1(k)$ and $h_0(k)$, governs everything that follows.

Iteration of the lowpass filter leads to the scaling function $\phi(t)$, which solves the two-scale equation (the most important equation)

$$\phi(t) = \sum h_0(k)\phi(2t - k).$$

The wavelets come from $\phi(t)$ by the highpass filter:

$$w(t) = \sum h_1(k)\phi(2t - k).$$

At every step of this theory we have convolution (which is the filter) and rescaling (n and t become $2n$ and $2t$). The approximation A_j at each level j is given by the scaling function. The details D_j are given by the wavelets. The key idea of *multiresolution* is this analysis of the signal into different scales and its reconstruction from the sum $D_1 + D_2 + D_3 + D_4 + D_5 + A_5$.

This new book by Charles Chui concentrates above all on *spline wavelets*. Those have special coefficients and special smoothness and symmetry. They approach Gaussians and modulated Gaussians as their degree increases. Therefore they come close to the optimal constant ($\frac{1}{2}$) in Heisenberg's Uncertainty Principle, which limits the sharpness of time-frequency windows.

The author has very wide experience in the practical applications of spline wavelets, and he writes about that too: *data compression* and *feature detection* and numerical solution of *integral equations*. The awkward technical problems of boundary wavelets have explicit solutions for splines. (The present writer has learned that when students come to ask advice about their theses, the problem is always at the boundary.) Image processing must deal with those problems.

Speaking of images, I hope you will appreciate the figures in this monograph. They convey lots of information. And no wonder: the old adage that 1 figure $= 10^3$ words is an enormous understatement. If we count bits, the ratio is much higher. The conclusion to draw is that figures are valuable, in fact absolutely necessary, but they have to be compressed. This is one of the principal applications, and achievements, of wavelets.

I welcome this new book from SIAM.

Gilbert Strang
Massachusetts Institute of Technology
gs@math.mit.edu

Preface

This writing is based on the lectures I gave at the SIAM one-day tutorial on "Wavelets and Applications" on July 11, the day before the 1993 SIAM Annual Meeting in Philadelphia. In preparation for these lectures, I focused on one theme among the multidisciplinary aspects of the fundamental concepts, theory, mathematical methods, and algorithms of the rapidly developing subject of wavelets. I chose signal processing as the main theme to help unify my presentation for this audience with a diverse background. The reasons for this choice were, first, I wanted to explore the notion of the (integral) wavelet transform, originally introduced by the geophysicist Jean Morlet for studying seismic waves, and, second, I had acquired a very strong interest in signal processing through research, guiding my electrical engineering and computer science students at Texas A&M University, and collaborating with several industrial companies.

To emphasize the importance of using wavelet analysis as a mathematical tool to understand and solve various problems in signal processing, I decided to sacrifice the mathematical generality, elegance, and abstraction. Instead, I outlined the wavelet analysis theory and its implications and included several illustrated examples. It was my goal that in so doing even an audience with undergraduate training in science or engineering could benefit from exposure to these lectures.

My goal agrees with the objectives of this new SIAM series edited by Joseph Flaherty. I would like to express my gratitude to Vickie Kearn and Susan Ciambrano of SIAM for bringing this series to my attention and for their encouragement and assistance in promptly publishing this volume. I would also like to thank Joe for his enthusiasm for this project as well as his cooperation throughout the reviewing process.

This is not a book on signal processing. Furthermore, it is not a comprehensive writing on the rapidly developing field of wavelet analysis. On the contrary, it is an elementary treatise of the most basic concepts, techniques, and computational algorithms of the wavelet transform, as well as the implications of this transform to the understanding and solution of various problems in signal processing. We give a fairly complete and concise treatment of the elements of wavelets pertinent to signal processing applications, including various motivations of the wavelet transform from the point of view of signal analysis, procedures for constructing some of the most popular wavelets, comparisons of these wavelets for multilevel bandpass filtering, as well as algorithms and computational schemes for computer implementation. To do all these in a small volume, we can neither afford to discuss the rich history of the subject of wavelets nor to elaborate on its rapidly growing and already vast literature. The interested reader is encouraged to look up the more complete bibliographies in the cited books and papers listed in the references. A short summary of the material presented in each chapter and some historical remarks along with pointers to further study are given in the section "Summary and Notes."

So, what are wavelets to the engineers who work with signals? Perhaps one of the most appropriate answers to this question is that a wavelet is a mathematical toolbox with two major components: analysis and synthesis. For analysis, a wavelet (or, more precisely, an analyzing wavelet) is an analog (convolution) bandpass filter with three distinct properties: (1) a scaling (or dilation) parameter is used to adjust the width of the frequency band along with the location of its center frequency; (2) a translation parameter is used to describe the position in time (or space) of the signal with frequency components in this band having the constant-Q property, which implies that this filter automatically zooms in to give precise locations of high-frequency changes and zooms out to give an accurate study of the low-frequency behavior; and (3) together with its corresponding scaling function as a lowpass filter, computational algorithms are most efficient. For synthesis, a wavelet (or, more precisely, a synthesizing wavelet) and perhaps a different scaling function as a smoothing function are used to model (or represent) a signal from its lowpass (background) features and bandpass (high-frequency) details. The pair of analyzing and synthesizing wavelets as well as the pair of lowpass and smoothing scaling functions are constructed so that the reconstructed signal remains the same as the input signal, as long as the background features and high-frequency details of the input signal are unchanged. This book emphasizes all of these aspects of wavelets. Various adaptations of this approach to take care of bounded intervals and the multidimensional settings, as well as to gain flexibility to achieve certain desirable properties in signal analysis, are also introduced in this writing. This monograph concludes with a chapter on various applications of the wavelet transform. Applications to signal analysis include detection of irregularities, feature extractions, compression of digital data (particularly of audio signals), still images, and video sequences. Other applications such as numerical solutions of integral equations will also be discussed.

One of the main objectives of this writing is to stimulate interactions among mathematicians, computer scientists, engineers, as well as biological and physical scientists. To meet this goal, I have adopted an informal writing style and selected only the elementary but basic topics related to the subject. It is my hope that this monograph can be appreciated by anyone with a very minimal knowledge of the Fourier approach.

While preparing the monograph, I received assistance from my former student J. Goswami in preparing most of the figures and tables in this volume and my post doc L. F. Zhong in going over with me all the details in the section on boundary spline wavelets with arbitrary knots in Chapter 6. To them, I am most grateful. The many helpful comments by the reviewers from various disciplines in mathematical sciences and engineering were instrumental in guiding me to write the first and final chapters, which hopefully will enhance the usefulness of this book, not only for signal analysis but also for other applications of wavelets. To them and to my colleague Álmos Elekes, who read over most of the manuscript and caught a number of typos, I am very grateful. I would also like to take this opportunity to acknowledge the following funding agencies and industrial companies for their support of my research and development activities on wavelet analysis and its various applications during the past seven years: National Science Foundation, Army Research Office, Naval Research Laboratories, Texas Coordinating Board of Higher Education, Houston Advanced Research Center, Texas Instruments, E–Systems, General Motors, and McDonnell Douglas Corp.

In putting this manuscript into its final form, I have again benefited from the excellent technical skill of Robin Campbell, who typed a portion of the manuscript, and my wife, Margaret, who assisted in the final details of the production of the manuscript. To both of them, I extend my most sincere thanks. Finally, I would like to thank SIAM for providing me with very skillful editorial assistance.

Charles K. Chui

Software

The following is a list of commercial and public domain wavelet software and/or sites known to the author. The inclusion of them does not represent the author's recommendation. Furthermore, the author is not responsible for the condition nor the accuracy of the software. This list is by no means complete, and the sites are subject to changes by the site owners.

The author would like to thank Margaret Chui and Veyis Nuri for their assistance in compiling this list.

1. *AccuPress* (wavelet image compression software), Aware, Inc.
    http://www.aware.com
2. *Adapted Waveform Analysis Library, v2.0*, Fast Mathematical Algorithms and Hardware Corporation, victor@math.wustl.edu
3. *AWA 3:Adapted Wavelet Analysis Library, v3*, Fast Mathematical Algorithms and Hardware Corporation, victor@math.wustl.edu
4. *epic* (pyramid wavelet coder). Contact Eero P. Simoncelli,
    eero@media.mit.edu
    whitechapel.media.mit.edu:/pub/epic.tar.Z
5. The Multirate Signal Processing Group, University of Wisconsin–Madison,
    http://saigon.ece.wisc.edu/~waveweb/QMF.html
6. *hcompress* (wavelet image compression),
    stsci.edu:/software/hcompress/hcompress.tar.Z
7. *LIFTPACK*. Contact Gabriel Fernandez, fernande@cs.sc.edu
    http://www.cs.sc.edu/fernande/liftpack/beta.html
8. *Orthonormal bases of compactly supported wavelets*, I. Daubechies. Journal of Applied and Computational Harmonic Analysis,
    http://wuarchive.wustl.edu/~acha
9. *rice-wlet* (wavelet software),
    cml.rice.edu:/pub/dsp/software/rice-wlet-tools.tar.Z
10. *SADAM*. Contact Fionn Murtagh,
    fmurtagh@cdsxb6.u-strasbr.fr
11. *scalable* (two- and three-dimensional subband transformation),
    scalable@robotics.eecs.berkeley.edu

robotics.eecs.berkeley.edu:/pub/multimedia/scalable2.tar.Z

12. *Self-diffusion maps from wavelet de-noised NMR images*, G. Sarty and E. Kendall, Journal of Magnetic Resonance, Series B, 111(1996), pp. 50–60. http://maya.usask.ca/sarty/sarty01.tar.Z

13. *The wavelet-based synthesis for the fractional Brownian motion proposed by F. Sellan and Y. Meyer: Remarks and fast implementation*, by P. Abry and F. Sellan. Journal of Applied and Computational Harmonic Analysis, http://wuarchive.wustl.edu/~acha

14. *WavBox Software*, WBTS. http://www.wavbox.com

15. *WaveLab*.701, http://playfair.stanford.edu/~waveweb

16. *Wavelet Explorer*, Wolfram Research. http://www.wolfram.com/wsn

17. *Wavelets in a Box*, Academic Press, Inc. Contact Charles Glaser, cbglaser@aol.com

18. *Wavelet Toolbox*, Cambridge University Technical Services Ltd. Contact David Newland, den@eng.cam.ac.uk

19. *wavethresh* (wavelet software for the language S), gpn@maths.bath.uk gdr.bath.ac.uk:/pub/masgpn/wavethresh2.2.Z

20. *wvlt* (package of wavelet transform routines in C). Contact Bob Lewis, bobl@cs.ubc.ca
http://www.cs.ubc.ca/nest/imager/contributions/bobl/wvlt/top.html

21. ftp://daisy/uwaterloo.ca/pub/maple/S.3/share/daub

22. ftp://info.mcs.anl.gov/pub/W-transform

23. ftp://mu.ceremade.dauphine.fr/pub/software

24. ftp://pandemonium.physics.missouri.edu/pub/wavelets

25. http://http.hg.eso.org/midas-info

26. http://iaks-www.ira.uka.de/iaks-beth/wavelet/software

27. http://jazz.rice.edu/software/RWT

28. http://summus.com

29. http://www.amara.com/wwbdev/wwbdev.html

30. http://www.atinternet.fr/image

31. http://www.c3.lanl.gov/~cjhamil/wavelets/main.html

32. http://www.cis.upenn.edu/~eero/epic.html

33. http://www.dfw.net/~ncody

34. http://www.infinop.com

35. http://www.intergalact.com/macwavelets/macwavelets.html

36. http://www.harc.edu/HARCC.html

37. http://www.math.yale.edu/pub/wavelets/software

38. http://www.mathworks.com/wavelet.html

39. ftp://www.pd.uwa.edu.au/pub/wavelets

40. http://www.swin.edu.au/chem/bio/s+code/wpacfrac1.htm

41. http://www.statci.com/wavelets.html

42. http://www.tsc.uvigo.es/~wavelets/uni-wave.html

43. http://www.vni.com

Notation

Notation	First appearance on page

CHAPTER 1

What are wavelets?

The term "wavelets" has a very broad meaning, ranging from singular integral operators of the Calderón type in harmonic analysis to subband coding algorithms in signal processing, from coherent states in quantum physics to spline analysis in approximation theory, from multiresolution transform in computer vision to a multilevel approach in the numerical solution of partial differential equations, and so on. To those of us who are interested in the understanding and analysis of signals, however, wavelets may be considered to be mathematical tools for waveform representations and segmentations, time-frequency analysis, and fast algorithms for easy implementation. These aspects of wavelets are the subject of this book. Since our primary goal is to reach the general audience in the science and engineering communities, we will focus on only the elementary and basic materials. Hence, current topics of interest such as localized cosine bases, multiwavelets, frames, matrix-dilated wavelet transforms, etc., are beyond the scope of this book and will not be covered.

1.1. Waveform modeling and segmentation.

The standard mathematical tool for representing signals, particularly those that are periodic, is a Fourier (or trigonometric) series. A finite sum (or truncation) of such an infinite series provides a mathematical model of the signal. Of course, a better model is usually achieved by increasing the number of terms in this trigonometric polynomial. In Figure 1.1, the original signal (on top) is represented by finite sums of its trigonometric series expansion with an increasing number of terms toward the bottom. Observe that even with six terms as shown at the bottom of Figure 1.1, we have a fairly good representation of the original signal. These trigonometric models are formulated by taking linear combinations of the basis functions shown in Figure 1.2.

On the other hand, the signal shown in Figure 1.3 requires many more terms of its trigonometric series expansion to achieve a reasonable model. Using the Walsh basis shown in Figure 1.4, however, only a few terms are needed to model the (piecewise constant) original signal. In Figure 1.5, an increasing number of terms of the Walsh series expansion is shown, with the bottom one requiring only six terms to model the original signal.

1

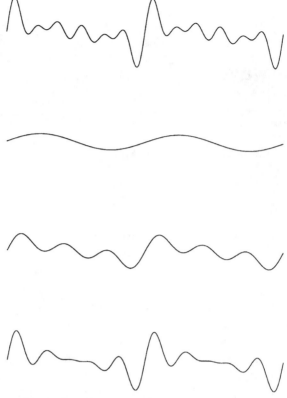

FIG. 1.1.   *Original signal and its two-, four- and six-term Fourier expansions.*

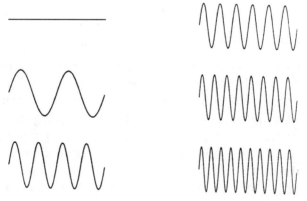

FIG. 1.2.   *Trigonometric basis functions.*

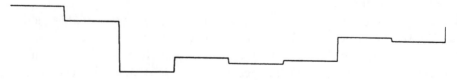

FIG. 1.3.   *Original signal.*

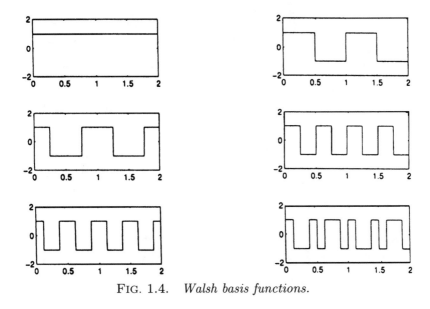

FIG. 1.4.   *Walsh basis functions.*

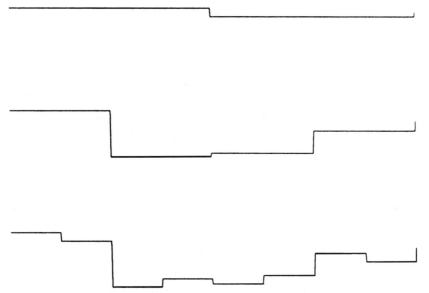

FIG. 1.5.   *Model of original signal using the two-, four- and six-term Walsh series.*

The above two examples illustrate the importance of choosing an appropriate basis for modeling a given signal for the purpose of signal processing. There is, however, a fundamental difference between the Walsh basis shown in Figure 1.4 and the trigonometric basis shown in Figure 1.2. Being piecewise constant, each of the Walsh basis functions can be conveniently segmented to gain both flexibility and a certain very rich algebraic structure to be described below. For instance, each of the two functions on the top of Figure 1.4 can be written as the sum of two "constant pieces" as shown in Figures 1.6 and 1.7,

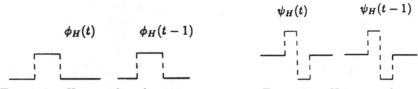

FIG. 1.6.   *Haar scaling functions.*   FIG. 1.7.   *Haar wavelets.*

respectively. In fact, the first Walsh basis function is nothing but the 2-periodic extension of the sum of the two constant pieces, $\phi_H(t)$ and its integer translate $\phi_H(t-1)$, shown in Figure 1.6, while the second Walsh basis function is the 2-periodic extension of the two constant pieces $\psi_H(t)$ and its integer translate $\psi_H(t-1)$ shown in Figure 1.7. Furthermore, the other Walsh basis functions are also generated by periodic extensions of linear combinations of scales by 2^{-j} and translates by $\frac{k}{2^j}$ of $\phi_H(t)$ and $\psi_H(t)$ for certain appropriate integers j and k, namely,

$$\phi_H(2^j t - k) \quad \text{and} \quad \psi_H(2^j t - k), \quad j, k \in \mathbb{Z}. \tag{1.1.1}$$

Here and throughout this book, the notation $\mathbb{Z} = \{\ldots, -1, 0, 1, \ldots\}$ for the set of all integers will be used.

In other words, the two functions $\phi_H(t)$ and $\psi_H(t)$ "generate" the entire Walsh basis; moreover, only ± 1 are used as coefficients in generating the linear combinations. Hence, segmentation of the Walsh basis by using the two basic functions $\phi_H(t)$ and $\psi_H(t)$, shown in the left of Figures 1.6 and 1.7, respectively, gives us the flexibility to use arbitrary constants, instead of only ± 1, as coefficients in taking linear combinations of the two scale-translation families in (1.1.1), yielding the function spaces

$$V_{H,j} = \left\{ \sum_k c_{j,k} \phi_H(2^j t - k) : \sum_k |c_{j,k}|^2 < \infty \right\} \tag{1.1.2}$$

and

$$W_{H,j} = \left\{ \sum_k d_{j,k} \psi_H(2^j t - k) : \sum_k |d_{j,k}|^2 < \infty \right\}, \tag{1.1.3}$$

where, for each integer j, the summations in the definition of $V_{H,j}$ and $W_{H,j}$ are taken over all bi-infinite square-summable sequences of real numbers. As a consequence, we do not have to restrict ourselves to periodic functions but, instead, have the freedom to extend to the space

$$L^2 := L^2(-\infty, \infty) \tag{1.1.4}$$

of square-integrable (or finite-energy) functions (or signals) $f(t)$ with

$$\|f\| = \left\{ \int_{\infty}^{\infty} |f(t)|^2 dt \right\}^{1/2} < \infty. \qquad (1.1.5)$$

The notation $\|\cdot\|$ in (1.1.5), called the L^2-norm, is used to measure the "energy" of functions (or signals) in L^2.

The structures of the function spaces $V_{H,j}$ and $W_{H,j}$ as vector subspaces of L^2 are very rich. First, from the definition in (1.1.2), we see that, for each fixed $j \in \mathbb{Z}$, the family

$$\phi_{H;j,k}(t) := 2^{j/2} \phi_H(2^j t - k), \quad k \in \mathbb{Z} \qquad (1.1.6)$$

is a "basis" of $V_{H,j}$. In fact, since the supports of $\phi_{H;j,k}(t)$ and $\phi_{H;j,\ell}(t)$ do not overlap whenever $k \neq \ell$, this basis is orthogonal. Hence, the normalization by the multiplicative constant $2^{j/2}$ in (1.1.6) implies that $\{\phi_{H;j,k}\}, k \in \mathbb{Z}$ is an orthonormal basis of $V_{H,j}$. Similarly, it is clear from (1.1.3) that for each fixed j the family

$$\psi_{H;j,k}(t) := 2^{j/2} \psi_H(2^j t - k), \quad k \in \mathbb{Z} \qquad (1.1.7)$$

is an orthonormal basis of $W_{H,j}$.

Two other important properties are that, first, the subspaces $V_{H,j}$, $j \in \mathbb{Z}$, of L^2 are nested in the sense that

$$\cdots \subset V_{H,-1} \subset V_{H,0} \subset V_{H,1} \subset V_{H,2} \subset \cdots \qquad (1.1.8)$$

and, second, for each $j \in \mathbb{Z}$,

$$\begin{cases} V_{H,j+1} = V_{H,j} + W_{H,j} \quad \text{and} \\ V_{H,j} \perp W_{H,j}, \end{cases} \qquad (1.1.9)$$

where the notation $\perp$ means that the two function spaces are orthogonal. To see that (1.1.8) holds, we simply observe that as we go up the ladder from $V_{H,j}$ to $V_{H,j+1}$ in (1.1.8), we just chop each constant piece into two pieces to gain additional freedom. Such freedom is important because it allows us to approximate every function $f(t)$ in L^2 as closely as we wish by functions in $V_{H,j}$, for a sufficiently large integer j, by using the L^2-norm measurement introduced in (1.1.5). This is similar to the approximation of a Riemann integral by a Riemann sum, although the mathematical argument is slightly more sophisticated. Hence, the nested structure of the subspaces $V_{H,j}$ in (1.1.8) can be formulated even more precisely as follows:

$$\cdots \subset V_{H,-1} \subset V_{H,0} \subset V_{H,1} \subset V_{H,2} \subset \cdots \rightarrow L^2. \qquad (1.1.10)$$

To see that (1.1.9) holds, let us first observe that it is sufficient to verify (1.1.9) for $j = 0$. Indeed, substitution of t by $2^j t$ then gives (1.1.9) for any $j \in \mathbb{Z}$. So for $j = 0$ the set identity in (1.1.9) is best justified by appealing to Figure 1.8, which says that

$$\begin{cases} \phi_H(2t) = \dfrac{1}{2}\phi_H(t) + \dfrac{1}{2}\psi_H(t), \\[2mm] \phi_H(2t - 1) = \dfrac{1}{2}\phi_H(t) - \dfrac{1}{2}\psi_H(t). \end{cases} \qquad (1.1.11)$$

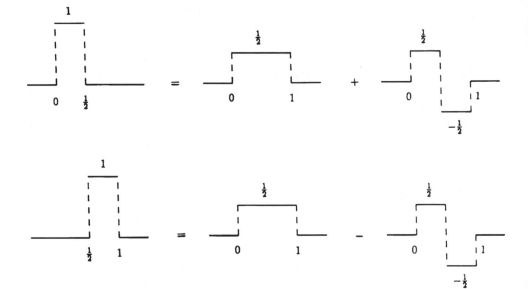

FIG. 1.8.   *Decomposition relations.*

Hence, by setting

$$\begin{cases} a_0 = a_1 = \dfrac{1}{2} \quad \text{and} \quad a_k = 0 \;\text{ for }\; k \neq 0, 1, \\[2mm] b_0 = \dfrac{1}{2}, \quad b_1 = -\dfrac{1}{2}, \quad \text{and} \quad b_k = 0 \;\text{ for }\; k \neq 0, 1, \end{cases} \qquad (1.1.12)$$

the two identities in (1.1.11) can be reformulated into a single identity

$$\phi_H(2t - \ell) = \sum_k \{a_{\ell-2k}\, \phi_H(t - k) + b_{\ell-2k}\, \psi_H(t - k)\} \qquad (1.1.13)$$

for $\ell = 0, 1$. The advantage of the formulation in (1.1.13) over that in (1.1.11) is that (1.1.13) holds not only for $\ell = 0, 1$ but also for all integers ℓ. Indeed, if ℓ is even, say $\ell = 2n$, then (1.1.13) is the same as the first identity in (1.1.11) with t replaced by $t - n$, and, similarly, $\ell = 2n + 1$ in (1.1.13) is the same as the second identity in (1.1.11) with t replaced by $t - n$.

Now, recalling from the definition of $V_{H,1}$ in (1.1.2) with $j = 1$ that the left-hand side of (1.1.13), $\ell \in \mathbb{Z}$, generates all of $V_{H,1}$, we see from the formulation of the right-hand side of (1.1.13) that, indeed, we have $V_{H,1} = V_{H,0} + W_{H,0}$. The orthogonality between $V_{H,0}$ and $W_{H,0}$ should be clear from Figures 1.6–1.7.

For completeness, we mention that the nested property of the subspaces $V_{H,j}$ and the inclusion property $W_{H,j} \subset V_{H,j+1}$ are governed by the identities

$$\begin{cases} \phi_H(t) = \sum_k p_k \phi_H(2t - k), \\ \psi_H(t) = \sum_k q_k \phi_H(2t - k), \end{cases} \tag{1.1.14}$$

respectively, with

$$\begin{cases} p_0 = p_1 = 1 \quad \text{and} \quad p_k = 0 \quad \text{for } k \neq 0, 1, \\ q_0 = 1, \quad q_1 = -1, \quad \text{and} \quad q_k = 0 \quad \text{for } k \neq 0, 1. \end{cases} \tag{1.1.15}$$

This is demonstrated by Figure 1.9.

The identities given in (1.1.14) are called two-scale relations, since they relate the generating scale spaces $V_{H,0}$ and $W_{H,0}$ to the first scale space $V_{H,1}$. For obvious reasons, $\phi_H(t)$ is then called a scaling function, and for certain reasons that will be clear later, $\psi_H(t)$ is called a *wavelet*. The subscript H stands for "Haar." Hence, $\phi_H(t)$ and $\psi_H(t)$ are called *Haar scaling function* and *Haar wavelet*, respectively.

Let us return to the trigonometric basis shown in Figure 1.2 and ask ourselves if an analogous segmentation would lead to the rich algebraic structures in (1.1.9)–(1.1.10) as partially described by (1.1.13)–(1.1.14). There are really not too many ways to localize the sine functions. Perhaps the most natural way is to consider the two segmentations as shown in Figure 1.10, where $\phi_t(t) = (\sin \pi t)\chi_{[0,1]}(t)$ and $\psi_t(t) = (\sin 2\pi t)\chi_{[0,1]}(t)$. Here and throughout, the notation $\chi_A(t)$ is used for the "characteristic function" of a set A, meaning that $\chi_A(t) = 1$ if $t \in A$ and $\chi_A(t) = 0$ otherwise.

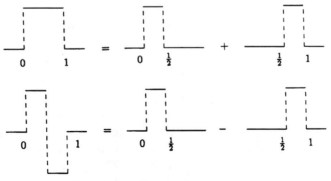

FIG. 1.9. *Two-scale relations.*

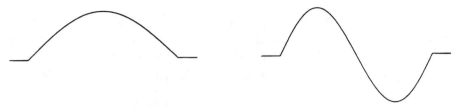

FIG. 1.10. *Truncated sine functions.*

However, although certain orthogonality relations still hold, none of the other algebraic properties (1.1.9)–(1.1.10) enjoyed by the Haar scaling function $\phi_H(t)$ and Haar wavelet $\psi_H(t)$ are valid for the segmented sinusoidal functions $\phi_t(t)$ and $\psi_t(t)$. For instance, there do *not* exist sequences of real numbers p_k and q_k that give two-scale relations

$$\phi_t(t) = \sum_k p_k \phi_t(2t - k)$$

and

$$\psi_t(t) = \sum_k q_k \phi_t(2t - k).$$

1.2. Time-frequency analysis.

In the above section, we have used the Haar scaling function $\phi_H(t)$ and Haar wavelet $\psi_H(t)$ to demonstrate the flexibility in signal representation by using a local basis, say $\{\phi_{H;j,k}(t) : k \in \mathbb{Z}\}$ in (1.1.6), of some subspace $V_{H,j}$ as defined in (1.1.2) of the nested sequence of subspaces in (1.1.10). We have also identified the orthogonality structure (1.1.9) enjoyed by $\phi_H(t)$ and $\psi_H(t)$. This structure, as we will discuss in this section, allows the Haar wavelet $\psi_H(t)$ to analyze the signal that has been represented by using the Haar scaling function $\phi_H(t)$. In addition, as we will see in the next section, the decomposition relation (1.1.13) gives rise to an efficient (in fact, real-time) algorithm for decomposing the signal for the purpose of analyzing the "time-frequency" content of the signal with the "zoom-in" and "zoom-out" capability. Similarly, the identities in (1.1.14), which we will call "two-scale relations" of $\phi_H(t)$ and $\psi_H(t)$, provide an efficient algorithm for perfectly reconstructing the signal after it has been analyzed. Unfortunately, with the exception of signals that are supposed to be piecewise constant, the Haar scaling function $\phi_H(t)$ is not effective in signal representation (or modeling), and the Haar wavelet $\psi_H(t)$ whose Fourier transform is not well localized (having a very slow rate of decay on the frequency axis) is not suitable for local time-frequency analysis. One of the main objectives of this book is to introduce to the reader, who may not be very familiar with the subject of wavelets, other scaling functions for signal modeling and their corresponding wavelets for effective time-frequency analysis. Since this first chapter is intended to give only a snapshot of the significance of wavelet analysis in signal processing, we assume that a desirable pair of scaling function $\phi(t)$ and wavelet $\psi(t)$ have been constructed. In order to give ourselves more freedom in constructing more desirable $\phi(t)$ and $\psi(t)$, we will not even impose any orthogonality conditions on $\phi(t)$ and $\psi(t)$ such as the orthogonality among the integer translates $\phi_H(t-k)$ and $\psi_H(t-k)$ of the Haar scaling function and Haar wavelet or the orthogonality between the subspaces $V_{H,j}$ and $W_{H,j}$ in (1.1.9) generated by $\phi_H(t)$ and $\psi_H(t)$. All we need is a "dual wavelet" $\widetilde{\psi}(t)$, which is orthogonal to the space V_0 generated by $\phi(t)$.

More precisely, let us assume that we have a scaling function $\phi(t)$ that generates a nested sequence of subspaces $\{V_j\}$ of the finite-energy space L^2 in the same way as the Haar scaling function does in (1.1.2) and (1.1.10); namely,

$$\cdots \subset V_{-1} \subset V_0 \subset V_1 \subset V_2 \subset \cdots \rightarrow L^2, \tag{1.2.1}$$

where

$$V_j = \mathrm{clos}_{L^2} \langle \phi(2^j t - k) : k \in \mathbb{Z} \rangle. \tag{1.2.2}$$

Here and throughout this book, since $\phi(t)$ may not have compact support (or finite time duration) as $\phi_H(t)$, we use the alternate standard notations $\langle\ \rangle$ and "L^2-closure" to indicate that V_j is the set of all possible finite linear combinations of $\phi(2^j t - k)$, with fixed j, as well as all their limits under the L^2-norm, respectively. In addition, let us assume that we have a pair of (dual) wavelets $\psi(t)$ and $\widetilde{\psi}(t)$ in L^2 with the properties that, first, the two families

$$\psi_{j,k}(t) := 2^{j/2}\psi(2^j t - k), \quad j, k \in \mathbb{Z}, \tag{1.2.3}$$

and

$$\widetilde{\psi}_{j,k}(t) := 2^{j/2}\widetilde{\psi}(2^j t - k), \quad j, k \in \mathbb{Z} \tag{1.2.4}$$

are biorthogonal in the sense of

$$\langle \psi_{j,k}(t), \widetilde{\psi}_{\ell,m}(t) \rangle := \int_{-\infty}^{\infty} \psi_{j,k}(t)\overline{\widetilde{\psi}_{\ell,m}(t)}dt = \delta_{j,\ell}\delta_{k,m}, \quad j, k, \ell, m, \in \mathbb{Z} \tag{1.2.5}$$

(where $\delta_{j,k}$ is the usual Kronecker delta symbol) and are "stable" (i.e., they are both Riesz bases of L^2, to be defined precisely in Chapters 3 and 5); second, the spaces

$$W_j := \mathrm{clos}_{L^2} \langle \psi(2^j t - k) : k \in \mathbb{Z} \rangle \tag{1.2.6}$$

generated by $\psi(t)$ are complementary subspaces of the nested sequence $\{V_j\}$, i.e.,

$$\begin{cases} V_{j+1} = V_j + W_j, \\ V_j \cap W_j = \{0\}, \quad j \in \mathbb{Z}; \end{cases} \tag{1.2.7}$$

and, finally, the dual $\widetilde{\psi}(t)$ of $\psi(t)$ is orthogonal to V_0, namely,

$$\langle \phi(t-k), \widetilde{\psi}(t) \rangle = 0, \quad k \in \mathbb{Z}. \tag{1.2.8}$$

In (1.2.5) and throughout this book, the bar over a function (or complex number) denotes complex conjugation. We remark that the "direct sum" decomposition in (1.2.7) is weaker than the orthogonal decomposition in (1.1.9) and that since the Haar wavelet $\psi_H(t)$ generates an orthonormal family $\psi_{H;j,k}(t)$, $j, k \in \mathbb{Z}$, $\psi_H(t)$ is already stable, and its dual is just itself, so that (1.2.8) is satisfied by $\widetilde{\psi}_H(t) = \psi_H(t)$. A more careful study of the structure (1.2.5) and (1.2.7)–(1.2.8) will be given in Chapter 5.

We are now in a position to discuss how the mathematical formulation in (1.2.1), (1.2.5), and (1.2.7)–(1.2.8) can be applied to analyze and synthesize signals. Their corresponding fast algorithms will be briefly illustrated in the next section and studied in greater detail in Chapter 6.

Let $f(t)$ be any finite-energy signal to be analyzed. Our first step is to represent $f(t)$ by a function $f_n(t)$ from V_n for some sufficiently large n. Since we will analyze $f_n(t)$ instead of $f(t)$ and reconstruct $f_n(t)$ after it has been analyzed, the "modeling" of $f(t)$ by $f_n(t)$ certainly deserves some care. In most applications, since only discrete information on the signal $f(t)$ is available, and this information may even be noisy, modeling of $f(t)$ by $f_n(t) \in V_n$ is unavoidable anyway. In this direction, it is worth mentioning that a suitable choice of the scaling function $\phi(t)$ (and consequently the subspaces V_j it generates) depends on the "shape" of signal $f(t)$ itself. This concept of signal matching has already been discussed in the previous section. Next, in view of the nested property (1.2.1) and the direct-sum and decomposition structures in (1.2.7), we see that there is a unique decomposition of the model $f_n(t)$ in terms of its components $g_j(t) \in W_j$ and "DC" term $f_{n-m}(t) \in V_{n-m}$; namely,

$$f_n(t) = g_{n-1}(t) + \cdots + g_{n-m}(t) + f_{n-m}(t), \tag{1.2.9}$$

where $m > 0$ is as large as desired. In Chapter 2, we will see that any wavelet is a "bandpass filter." In fact, by introducing an appropriate constant $c > 0$ (in frequency units, such as Hz), the mapping

$$a \longmapsto w = \frac{c}{a} \tag{1.2.10}$$

from scale a to frequency w allows us to think of W_j as the space that governs the jth octave (or frequency) band of the frequency domain. In other words, the jth component

$$g_j(t) =: \sum_k \widehat{d}_{j,k} \psi_{j,k}(t)$$

$$= \sum_k \widehat{d}_{j,k} 2^{j/2} \psi(2^j t - k) \tag{1.2.11}$$

of the signal $f_n(t)$ in (1.2.9) "lives" in the jth octave band (which is well justified from the "frequency coefficient" 2^j of t in the series representation (1.2.11) of $g_j(t)$). Similarly, as we will see in the next chapter, any scaling function can be considered a "lowpass filter." Hence, with the scale-frequency transformation (1.2.10), the "bandwidth" of the subspace V_j may be considered as $c\, 2^j$. In other words, the DC component $f_{n-m}(t)$ of $f_n(t)$ in (1.2.9) has bandwidth $c\, 2^{n-m}$, and this width decreases to zero as m tends to infinity. This is why we say that $f_{n-m}(t)$ is the DC component of $f_n(t)$ in the decomposition (1.2.9).

Returning to the series representation (1.2.11) of the jth octave component $g_j(t)$ of the signal $f_n(t)$, we now demonstrate the importance of the biorthogonality property (1.2.5) and the orthogonality condition (1.2.8) of the dual wavelet $\widetilde{\psi}(t)$ in time-frequency analysis. For each j, $n - m \le j \le n - 1$, since

$$g_j(t) = f_n(t) - (g_{n-1}(t) + \cdots + g_{j+1}(t) + f_j(t)),$$

it follows from (1.2.5) and (1.2.8) that

$$\langle g_j(t), \widetilde{\psi}(2^j t - k)\rangle = \langle f_n(t), \widetilde{\psi}(2^j t - k)\rangle$$

for all $k \in \mathbb{Z}$. Hence, by the biorthogonal property (1.2.5) again, we have, from (1.2.11),

$$
\begin{aligned}
\widehat{d}_{j,k} &= \langle g_j(t), \widetilde{\psi}_{j,k}(t)\rangle = \langle f_n(t), \widetilde{\psi}_{j,k}(t)\rangle \\
&= 2^{j/2} \int_{-\infty}^{\infty} f_n(t)\overline{\widetilde{\psi}\left(\frac{t - \frac{k}{2^j}}{\frac{1}{2^j}}\right)}\, dt.
\end{aligned}
\tag{1.2.12}
$$

In the next chapter, we will introduce the concept of *integral wavelet transform* (IWT), also known as *continuous wavelet transform* (CWT), defined by

$$(W_{\widetilde{\psi}} f)(b, a) = \frac{1}{\sqrt{a}} \int_{-\infty}^{\infty} f(t)\overline{\widetilde{\psi}\left(\frac{t - b}{a}\right)}\, dt, \tag{1.2.13}$$

of finite-energy signals $f(t)$ using the "analyzing wavelet" $\widetilde{\psi}(t)$ as the bandpass filter. From (1.2.12) we see that the coefficients $\widehat{d}_{j,k}$ of the wavelet series representation (1.2.11) of the jth component $g_j(t)$ of the signal $f_n(t)$ are simply the IWT values of $f_n(t)$ at the time-scale location

$$(b, a) = \left(\frac{k}{2^j}, \frac{1}{2^j}\right) \tag{1.2.14}$$

or, equivalently, the time-frequency location

$$(b, w) = \left(\frac{k}{2^j}, c2^j\right). \tag{1.2.15}$$

This doubly indexed sequence $\{\widehat{d}_{j,k}\}$ is called the *discrete wavelet transform* (DWT) of $f_n(t)$. Since $f_n(t)$ is supposed to be a good approximation of the original signal, $\{\widehat{d}_{j,k}\}$ is also called the DWT of $f(t)$.

If both $\widetilde{\psi}(x)$ and its Fourier transform $\widehat{\widetilde{\psi}}(\omega)$ are "well localized" (a notion that will be explored in some details in Chapter 2), then the DWT gives fairly precise information on the time-frequency content of the signal, with the indices j and k indicating the location near $\frac{k}{2^j}$ in the time domain at which the frequency of the signal is in the jth octave band. In addition, since the time location near $\frac{k}{2^j}$ becomes more accurate for large values of j (i.e., in high-frequency ranges), we see that the wavelet transform has the so-called zoom-in capability. Of course the accuracy of the time position is better measured in the form of the size of a "window," the definition of which will be made precise in Chapter 2. With this concept of time window, we see that the time window of the IWT narrows at high-frequency ranges and widens at low-frequency ranges, and the mapping in (1.2.10) assures that this is done automatically. This capability of the IWT is exactly what is needed for analyzing nonstationary signals.

Observe that the mathematical formulation of the Fourier transform alone does not give rise to any localization capability. In Figure 1.12, we show the magnitude of the Fourier transform of the music data displayed in Figure 1.11. This musical piece contains not only very high frequency values, but, unfortunately, certain high-frequency noise is also mixed in with the music and occurs in the same frequency ranges. If a standard lowpass filter is applied to remove the noise, the high-frequency music content is removed as well. This is shown in Figures 1.13 and 1.14. The smooth curve in Figure 1.13 means that the high-frequency content of the music has been removed, and the music then sounds flat. For comparison, we plot the IWT of the same noisy music data in Figure 1.15, where the frequency axis of the time-frequency plane is matched with the frequency domain that displays the Fourier transform of the data. Observe that since the noise content has larger amplitude than the music itself on the same frequency ranges, it shows up much more prominently. In Figure 1.16, we also show the DWT of the same music data. The rectangular structure indicates that only discrete values of the IWT are obtained, but they

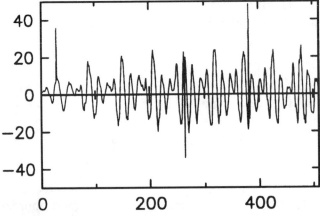

FIG. 1.11.   *Music data with high-frequency (pop) noise.*

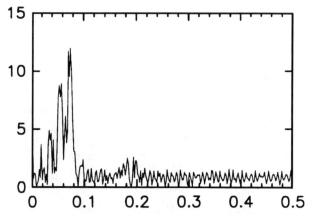

FIG. 1.12. *Magnitude of the Fourier transform of music data.*

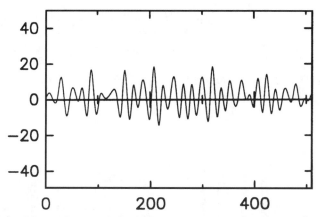

FIG. 1.13. *Music data with high-frequency noise (and music) removed.*

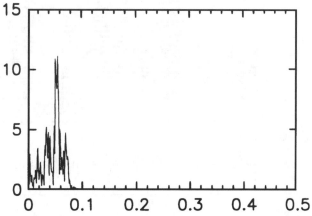

FIG. 1.14. *Fourier transform of "smoothed" music data.*

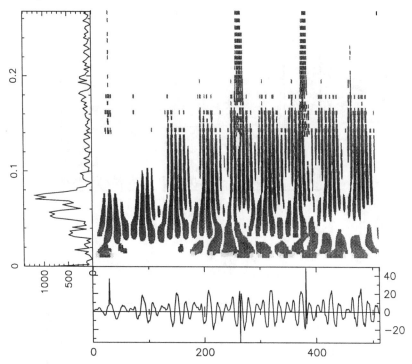

FIG. 1.15.   *Integral (or continuous) wavelet transform of music data.*

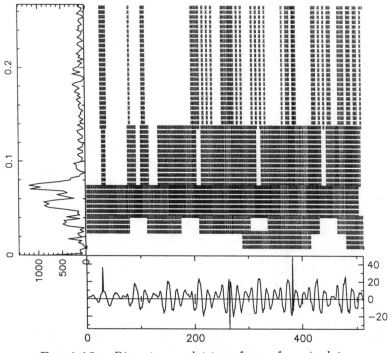

FIG. 1.16.   *Discrete wavelet transform of music data.*

are used to fill up the time-frequency plane. Observe that even this sparse set of data detects most of the high frequency noise content.

In Figure 1.17, we demonstrate the concept of wavelet decomposition of (a longer piece of) the same music data modeled by $f_n(t)$. The signal $f_n(t)$ shown at the top left-hand corner is decomposed as the sum of the $f_{n-1}(t)$ and $g_{n-1}(t)$ components, shown in the second row. The third row is the decomposition of $f_{n-1}(t)$ as the sum of the $f_{n-2}(t)$ and $g_{n-2}(t)$, etc. Observe that the large amplitudes of the noise content in $g_{n-1}(t)$ and $g_{n-2}(t)$ are quite prominent. After truncating these values, we can reconstruct the signal as shown in Figure 1.18. Here, while the high-frequency music is unchanged, only the high-frequency noise has been removed.

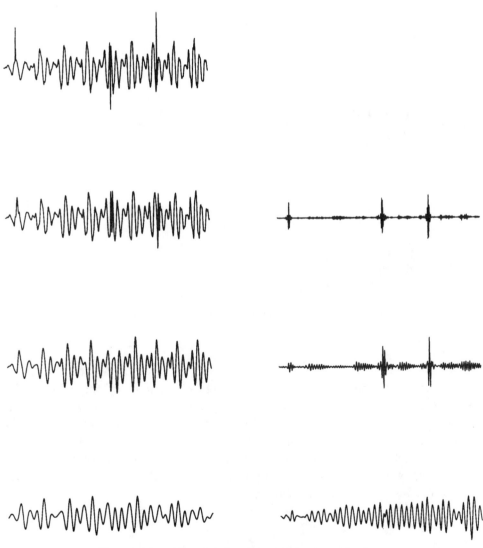

FIG. 1.17. *Wavelet decomposition of music data.*

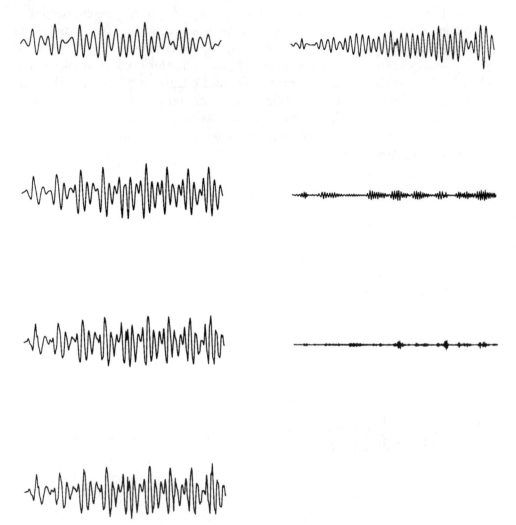

FIG. 1.18.   *Wavelet reconstruction of music data after removal of pop noise.*

As another example, a sinusoidal curve has been slightly perturbed and is represented by $f_n(t)$ in Figure 1.19. This perturbation shows up very clearly as $g_{n-1}(t)$ in Figure 1.20. The low-frequency content (of the pure sinusoidal signal) occurs in the DC component $f_{n-1}(t)$ but is not shown here. This example indicates the importance of the DWT in signal detection. In the next section, we will show that the computation of the DWT is even faster than that of the fast Fourier transform (FFT).

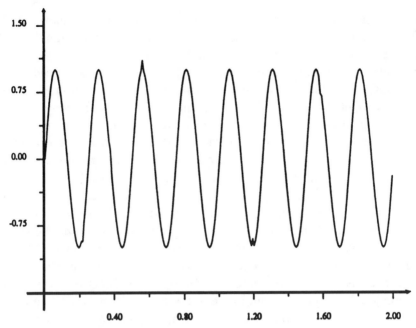

FIG. 1.19. *Perturbed sinusoidal curve.*

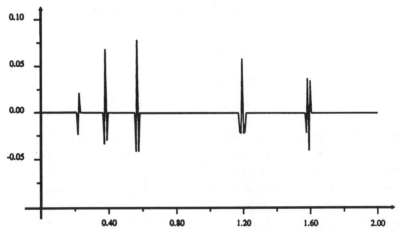

FIG. 1.20. *Perturbation magnified after sinusoidal curve is removed.*

1.3. Fast algorithms and filter banks.

The DWT introduced in the previous section is important not only because it gives local time-frequency information of a finite-energy (analog) signal but also because of its corresponding fast algorithms for both computational and implementational purposes. The aim of this section is only to give a very brief idea of how the basic fast algorithms work. Details of these and other related algorithms will be presented in Chapter 6.

As mentioned in the previous section, the wavelet transform (IWT or CWT) is defined for continuous-time functions (or analog signals). Hence, when discrete samples (or digital signals) are considered, they must first be mapped into the continuous-time domain. We may call this procedure "waveform matching." This preprocessing stage should not be completely ignored since the wavelet transform (including the DWT) has the so-called perfect reconstruction property. That is, whatever analog models are used as input signals, the same output signals are recovered, provided that no processing is applied to the wavelet (transform) components. Furthermore, the DWT information contained in the wavelet coefficients $\widehat{d}_{j,k}$, are values of the IWT of the analog model $f_n(t)$. In applications to data compression (see Chapter 7), for example, since redundancy is to be reduced with as little damage to the original data as possible, it is important to accurately analyze the signal and to be able to reconstruct the input data with very minimal loss.

The importance of signal representation (or waveform matching) by using scaling functions (other than simply the Haar scaling function) was mentioned in the previous section. It should be emphasized that even if the input signal is in analog form, it still requires remodeling by using an appropriate scaling function. The reason this is necessary is that the fast algorithms discussed below only apply to models given by some scaling function. More precisely, if $\phi(t)$ is the scaling function of choice, then by using a representation of the form

$$f_n(t) = \sum_k c_{n,k}\phi(2^n t - k), \tag{1.3.1}$$

fast algorithms can be applied to the coefficient sequence

$$\mathbf{c}_n = \{c_{n,k}\}, \quad k \in \mathbb{Z}, \tag{1.3.2}$$

instead of the original data. Here, the sequence $\mathbf{c}_n$ is chosen so that $f_n(t)$ is a desirable model of the original data (either in digital or analog formats). Although a larger value of n gives a better approximation, the value of this integer is limited by the physical constraints of the problem under consideration or by the desire to perform more effectively.

To describe the fast algorithms, we return to the structure of a nested sequence of subspaces:

$$\cdots \subset V_{-1} \subset V_0 \subset V_1 \subset V_2 \subset \cdots \to L^2 \tag{1.3.3}$$

generated by the scaling function $\phi(t)$ as in (1.2.1)–(1.2.2) and the complementary subspaces W_j, namely,

$$\begin{cases} V_{j+1} = V_j + W_j, \\ V_j \cap W_j = \{0\}, \end{cases} \tag{1.3.4}$$

generated by a corresponding wavelet $\psi(t)$ as in (1.2.6). The unique decomposition in (1.3.4) is governed by a pair of sequences $(\{a_k\}, \{b_k\})$ in the form

$$\phi(2t - \ell) = \sum_k \{a_{\ell-2k}\phi(t-k) + b_{\ell-2k}\psi(t-k)\}, \quad \ell \in \mathbb{Z}. \tag{1.3.5}$$

Here, although the relation only describes $V_1 = V_0 + W_0$, the same identity, by changing t to $2^j t$, describes $V_{j+1} = V_j + W_j$ for all integers j. On the other hand, the subspace structure

$$V_j \subset V_{j+1} \text{ and } W_j \subset V_{j+1}, \quad j \in \mathbb{Z} \tag{1.3.6}$$

is governed by another pair of sequences $(\{p_k\}, \{q_k\})$ in the form

$$\begin{cases} \phi(t) = \sum_k p_k \phi(2t - k), \\ \psi(t) = \sum_k q_k \phi(2t - k), \end{cases} \tag{1.3.7}$$

which describes (1.3.6) for all integers j when t is replaced by $2^j t$. The reader is reminded of the formulations (1.3.5) and (1.3.7) for the Haar setting in (1.1.13) and (1.1.14) with values of a_k, b_k, p_k, q_k given by (1.1.12) and (1.1.15).

With the two pairs of sequences, $\{a_k\}, \{b_k\}$ in (1.3.5) and $\{p_k\}, \{q_k\}$ in (1.3.7), we can write down precisely how the analog model $f_n(t)$ in (1.3.1) is decomposed as

$$f_n(t) = f_{n-1}(t) + g_{n-1}(t) \tag{1.3.8}$$

with $f_{n-1}(t) \in V_{n-1}$ and $g_{n-1}(t) \in W_{n-1}$ in the form of

$$\begin{cases} f_{n-1}(t) = \sum_k c_{n-1,k} \phi(2^{n-1}t - k), \\ g_{n-1}(t) = \sum_k d_{n-1,k} \psi(2^{n-1}t - k), \end{cases} \tag{1.3.9}$$

and how $f_n(t)$ is recovered from $f_{n-1}(t)$ and $g_{n-1}(t)$.

As mentioned above, in order to apply the fast algorithms, we must work with the coefficient sequence c_n in (1.3.2) instead of the analog model $f_n(t)$ in (1.3.1) it represents. Hence, the sequences

$$\mathbf{c}_{n-1} = \{c_{n-1,k}\} \text{ and } \mathbf{d}_{n-1} = \{d_{n-1,k}\} \tag{1.3.10}$$

in (1.3.9) that represent the components $f_{n-1}(t)$ and $g_{n-1}(t)$ of $f_n(t)$ will be used as output or input data sequences in the computational (or processing) algorithms.

A. Decomposition algorithm. *To find $f_{n-1}(t)$ and $g_{n-1}(t)$ from $f_n(t)$, compute*

$$\begin{cases} c_{n-1,k} = \sum_\ell a_{\ell-2k} c_{n,\ell}, \\ d_{n-1,k} = \sum_\ell b_{\ell-2k} c_{n,\ell}. \end{cases} \tag{1.3.11}$$

Observe that with the exception of sign changes in the indices of the filter sequences $\{a_k\}$ and $\{b_k\}$, the two (parallel) operations in (1.3.11) are really

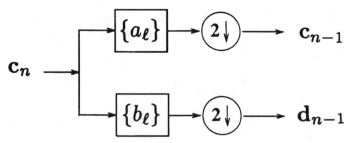

FIG. 1.21.   *Wavelet decomposition.*

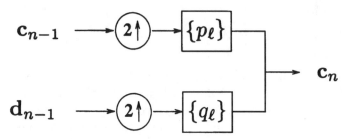

FIG. 1.22.   *Wavelet reconstruction.*

discrete convolutions. That is, the input sequence $\mathbf{c}_n = \{c_{n,\ell}\}$ is convolved with the filter sequences $\{a_{-\ell}\}$ and $\{b_{-\ell}\}$ in (1.3.5). The only difference between these operations and the ordinary convolution filter is that only the even-indexed output samples are kept (and relabeled: changing from $2k$ to k), while the odd-indexed output samples are discarded. This operation is called "downsampling" and is denoted by $2 \downarrow$. To describe the procedure (1.3.11) schematically, we refer to Figure 1.21.

B. Reconstruction algorithm. *To recover $f_n(t)$ from $f_{n-1}(t)$ and $g_{n-1}(t)$, compute*

$$c_{n,k} = \sum_\ell \{p_{k-2\ell}c_{n-1,\ell} + q_{k-2\ell}d_{n-1,\ell}\}. \tag{1.3.12}$$

Observe that if the index ℓ of the input sequences $c_{n-1,\ell}$ and $d_{n-1,\ell}$ were 2ℓ, then by considering the odd-indexed input data to be zero, the operation in (1.3.12) is precisely discrete convolution. The insertion of one zero between every two data samples is called "upsampling," and this operation is denoted by $2 \uparrow$. The schematic diagram for the procedure in (1.3.12) is shown in Figure 1.22.

If we put Figures 1.21 and Figure 1.22 together, we obtain a "filter bank" structure as shown in Figure 1.23.

Observe that the input sequence $\mathbf{c}_n$ is perfectly reconstructed. Filter banks in digital signal processing, particularly for subband coding, are actually more general. In the first place, we have only shown the "2-band" structure in Figure 1.23, while any number of bands may be used in subband coding. In the second place, the filter sequences $\{a_k\}$, $\{b_k\}$, $\{p_k\}$, and $\{q_k\}$ in subband coding are less restrictive than those in the above two algorithms. Recall that the first two sequences here govern the decomposition relation (1.3.5), while the other two

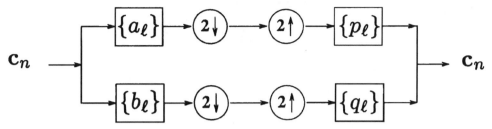

FIG. 1.23.   *Wavelet decomposition—reconstruction.*

sequences govern the subspace structure (1.3.6). In addition, our input and output sequences $\mathbf{c}_n, \mathbf{c}_{n-1}$, and $\mathbf{d}_{n-1}$ are not digital signals but digital representations (in the form of coefficient sequences) of certain analog models. In fact, these two algorithms are used to compute the DWT of an analog signal and to recover the analog signal from its DWT. Indeed, as shown in (1.2.12), with the exception of the normalization constant $2^{j/2}$, the output sequence $\mathbf{d}_{n-1}$ gives ($2^{-(n-1)/2}$ multiple of) the DWT of $f_n(t)$ at the time-scale locations $(\frac{k}{2^{n-1}}, \frac{1}{2^{n-1}}), k \in \mathbb{Z}$, by using the dual wavelet $\tilde{\psi}(t)$ as the analyzing wavelet as in (1.2.13). By repeated applications of the decomposition algorithm, Algorithm A, with input sequences $\mathbf{c}_{n-1}, \mathbf{c}_{n-2}, \ldots$, we can also compute the DWT of $f_n(t)$ at the time-scale locations

$$\left(\frac{k}{2^j}, \frac{1}{2^j}\right), \quad k \in \mathbb{Z}, \quad j = n-1, n-2, \ldots,$$

as mentioned in (1.2.14), again with the exception of the normalization constants $2^{j/2}$, $j \in \mathbb{Z}$. The only reason for dropping these constants in (1.3.1) and (1.3.9) is that we avoid multiplying the convolution operation by $\sqrt{2}$ in the decomposition algorithm (1.3.11) and that of the reconstruction algorithm (1.3.12) by $\frac{1}{\sqrt{2}}$.

CHAPTER 2

Time-Frequency Localization

To understand the impact of wavelets on science and technology, it is perhaps best to view wavelets as a time-frequency localization tool having at least two distinct features: the time-frequency window is flexible with the automatic zoom-in and zoom-out capability, and wavelet computational algorithms are usually efficient and simple to implement on the computer or processor. The objective of this chapter is to introduce the concept of time-frequency localization. The classical approach to analog filtering is reviewed in the first section. In section 2.2, the notion of *bandwidth* is generalized to that of root mean square (RMS) bandwidth through *windowing*. By introducing a phase parameter, this notion is further generalized to that of the *short-time Fourier transform* (STFT), in section 2.3. In section 2.4, the *integral wavelet transform* (IWT), also called the continuous wavelet transform (CWT), is introduced and compared with the STFT. To demonstrate the importance of the IWT, the final section of this chapter is devoted to a discussion of wavelet modeling of the cochlea of the human ear.

2.1. Analog filters.

An analog, or continuous-time, signal is a "measurable" function on the real line $\mathbb{R} = (-\infty, \infty)$, called the time domain. In practice, any analog signal is a continuous function with the exception of perhaps a countable number of jumps that occur on a discrete set without any finite accumulation points. Hence, for convenience, we will treat any analog signal as a "piecewise continuous" function, although the number of jump discontinuities of this function might be infinite. Almost all analog signals $f(t)$ of interest to the engineer have finite energy. By this we mean that $f(t)$ is square integrable; namely,

$$\int_{-\infty}^{\infty} |f(t)|^2 dt < \infty. \tag{2.1.1}$$

If $f(t)$ satisfies (2.1.1), we will write

$$f \in L^2 \quad \text{or} \quad f(t) \in L^2, \tag{2.1.2}$$

where $L^2 = L^2(-\infty, \infty)$, and we will use the "norm" notation

$$\|f\| = \left(\int_{-\infty}^{\infty} |f(t)|^2 dt \right)^{1/2} \tag{2.1.3}$$

to represent the square root of the total energy content of the signal $f(t)$.

The spectrum of a finite-energy signal $f(t)$ is described by its Fourier transform, defined by

$$\widehat{f}(\omega) = \int_{-\infty}^{\infty} e^{-j\omega t} f(t) dt, \tag{2.1.4}$$

where $j = \sqrt{-1}$ is the (unit) imaginary number, and ω will be called the frequency variable for convenience, although the frequency measurement is usually given by $\frac{\omega}{2\pi}$ (in terms of Hz). Hence, the signal $f(t)$ is called a *band-limited signal* if its Fourier transform $\widehat{f}(\omega)$ has compact support in the sense that

$$\widehat{f}(\omega) = 0 \quad \text{for} \quad |\omega| > \omega_0 \tag{2.1.5}$$

for some $\omega_0 > 0$. If ω_0 is the smallest value for which (2.1.5) holds, then the value

$$\omega_0 \tag{2.1.6}$$

is called the bandwidth of the signal $f(t)$.

Even if an analog signal $f(t)$ were not bandlimited, we can still change it to a bandlimited signal by what is called *ideal lowpass filtering*. For instance, to change $f(t)$ to a bandlimited signal $f_{\omega_0}(t)$ with bandwidth $\leq \omega_0$, we may simply consider

$$\widehat{f}_{\omega_0}(\omega) = \begin{cases} \widehat{f}(\omega) & \text{for} \quad |\omega| \leq \omega_0, \\ 0 & \text{for} \quad |\omega| > \omega_0 \end{cases} \tag{2.1.7}$$

and obtain $f_{\omega_0}(t)$ by the following *inverse Fourier transformation* operation:

$$f_{\omega_0}(t) = \frac{1}{2\pi} \int_{-\infty}^{\infty} e^{j\omega t} \widehat{f}_{\omega_0}(\omega) d\omega. \tag{2.1.8}$$

To better understand the formulation in (2.1.8), we need the definition of the characteristic function. For any measurable set A in $\mathbb{R}$, the function

$$\chi_A(t) = \begin{cases} 1 & \text{for} \quad t \in A, \\ 0 & \text{for} \quad t \notin A \end{cases} \tag{2.1.9}$$

is called the *characteristic function* of the set A. Hence, with ω as the variable of interest, if the characteristic function

$$\chi_{[-\omega_0, \omega_0]}(\omega)$$

is used as the *transfer function*, it provides an ideal lowpass filter; namely,

$$\widehat{f}_{\omega_0}(\omega) = \chi_{[-\omega_0, \omega_0]}(\omega) \widehat{f}(\omega), \tag{2.1.10}$$

which is simply a reformulation of (2.1.7). Hence, the bandlimited signal $f_{\omega_0}(t)$ obtained by using (2.1.8) is given by

$$f_{\omega_0}(t) = \frac{1}{2\pi} \int_{-\omega_0}^{\omega_0} e^{j\omega t} \widehat{f}(\omega)d\omega. \qquad (2.1.11)$$

Of course, if $f(t)$ is already a bandlimited signal with bandwidth not exceeding ω_0, then the lowpass filtering process (2.1.10) does not change the original signal $f(t)$; that is, $f_{\omega_0}(t) = f(t)$.

Let $\ell_{\omega_0}(t)$ be the inverse Fourier transform of the transfer function $\chi_{[-\omega_0,\omega_0]}(\omega)$. Then we have

$$\ell_{\omega_0}(t) = \frac{1}{2\pi} \int_{-\omega_0}^{\omega_0} e^{j\omega t}d\omega = \frac{\sin \omega_0 t}{\pi t}. \qquad (2.1.12)$$

This function will be called the *Shannon sampling function* (or, for $\omega_0 = \pi$, the *Shannon scaling function* to be discussed in some detail in the next chapter).

Analog filtering (or, more precisely, linear analog filtering) is defined by *time-domain convolution*. That is, if $h(t)$ is the filter function, then the input–output relation of this filter is given by

$$g(t) = (h * f)(t) = \int_{-\infty}^{\infty} h(x)f(t-x)dx = \int_{-\infty}^{\infty} f(x)h(t-x)dx. \qquad (2.1.13)$$

See Figure 2.1.

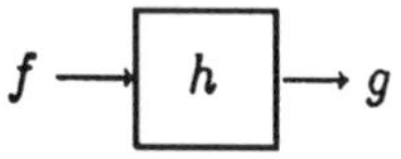

FIG. 2.1.   *Analog filtering (time-domain description).*

In the frequency domain, the filtering process (2.1.13) is described by

$$\widehat{g}(\omega) = H(\omega)\widehat{f}(\omega), \qquad (2.1.14)$$

where $H(\omega) = \widehat{h}(\omega)$ is the transfer function of the filter. Hence, in the frequency domain, filtering is accomplished simply by pointwise multiplication. This is shown in Figure 2.2.

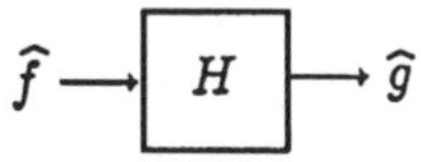

FIG. 2.2.   *Analog filtering (frequency-domain description).*

Returning to the mapping (2.1.10) of an analog signal $f(t)$ to a bandlimited signal $f_{\omega_0}(t)$, we have

$$f_{\omega_0}(t) = (\ell_{\omega_0} * f)(t) = \int_{-\infty}^{\infty} \ell_{\omega_0}(t-x)f(x)dx, \qquad (2.1.15)$$

with $\ell_{\omega_0}(t)$ as given by (2.1.12). In view of (2.1.13)–(2.1.14), the function $\ell_{\omega_0}(t)$, which has very slow decay, gives rise to the ideal lowpass filtering requirement (2.1.10). See Figure 2.3a–b.

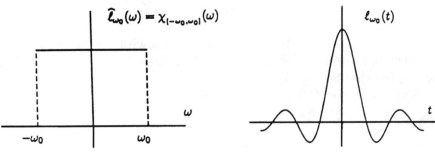

FIG. 2.3a.   *Ideal lowpass filter characteristic.*   FIG. 2.3b.   *Lowpass filter function.*

Let us consider an infinite sequence of positive numbers:

$$0 < \omega_0 < \omega_1 < \cdots < \omega_n < \cdots, \qquad \omega_n \to \infty.$$

In analogy to ideal lowpass filtering, we now discuss ideal bandpass filtering. This is accomplished by introducing the bandpass filter functions $b_{\omega_{n-1},\omega_n}(t)$ defined by

$$\widehat{b}_{\omega_{n-1},\omega_n}(\omega) = \chi_{[-\omega_n,-\omega_{n-1}]}(\omega) + \chi_{[\omega_{n-1},\omega_n]}(\omega) \qquad (2.1.16)$$

or, equivalently,

$$b_{\omega_{n-1},\omega_n}(t) = \frac{1}{\pi} \int_{\omega_{n-1}}^{\omega_n} \cos \omega t \; d\omega = \frac{\sin \omega_n t - \sin \omega_{n-1} t}{\pi t}. \qquad (2.1.17)$$

See Figure 2.4a–b.

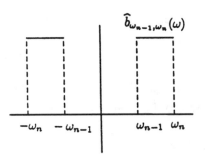
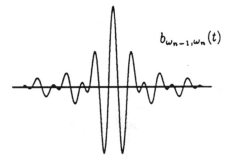

FIG. 2.4a.   *Ideal bandpass filter characteristic.*   FIG. 2.4b.   *Bandpass filter function.*

To stop the frequency content of an analog signal $f(t)$ outside the band $[\omega_{n-1}, \omega_n]$, one simply applies the bandpass filter $b_{\omega_{n-1},\omega_n}(t)$; namely,

$$\begin{cases} g_n(t) = (b_{\omega_{n-1},\omega_n} * f)(t) \quad \text{or} \\ \widehat{g}_n(\omega) = \widehat{b}_{\omega_{n-1},\omega_n}(\omega)\widehat{f}(\omega). \end{cases} \qquad (2.1.18)$$

Let $0 < \omega_0 < \cdots < \omega_N = \Omega$. If $f_\Omega(t)$ is a bandlimited signal with bandwidth $\leq \Omega$, then this signal can be decomposed as

$$f_\Omega(t) = f_{\Omega,\omega_0}(t) + g_{\Omega,1}(t) + \cdots + g_{\Omega,N}(t), \qquad (2.1.19)$$

where

$$\begin{cases} f_{\Omega,\omega_0} = (\ell_{\omega_0} * f_\Omega)(t) \quad \text{and} \\ g_{\Omega,n}(t) = (b_{\omega_{n-1},\omega_n} * f_\Omega)(t), \qquad n = 1, \ldots, N, \end{cases} \qquad (2.1.20)$$

with

$$\omega_N := \Omega. \qquad (2.1.21)$$

Observe that the component $f_{\Omega,\omega_0}(t)$, called the DC term, contains the low-frequency content of $f_\Omega(t)$, and each $g_{\Omega,n}(t)$ gives the precise information of $f(t)$ on the frequency band $[\omega_{n-1}, \omega_n]$, $n = 1, \ldots, N$.

2.2. RMS bandwidths.

Observe that each of the filter functions, ℓ_{ω_0}, $b_{\omega_0,\omega_1}, \ldots$, introduced in (2.1.12) and (2.1.17), provides an ideal frequency localization window by means of the convolution integral defined in (2.1.13). Unfortunately, as we will see in this section, these so-called ideal filters give a very poor time localization of the signal. In other words, they cannot be applied to give very precise information on when (or where) a signal behaves differently at certain frequency ranges of special interest. For this reason, we need filters that do not have ideal filter characteristics.

Let us now introduce the notion of windowing. A function $h(t)$, with certain desirable properties to be described below, will be called a *window function* (or time-window function), and we will use the "inner product" operation as the windowing process; that is, to window a signal $f(t)$ "near $t = b$," we consider the integral transform

$$\int_{-\infty}^{\infty} f(t)\overline{h(t-b)}\, dt. \qquad (2.2.1)$$

Hence, if $h(t)$ is a real-valued even function (so that $h(-t) = \overline{h(t)}$), then the windowing process (2.2.1) agrees with the convolution operation (2.1.13). That is, a real-valued even window function can also be treated as a filter function.

For example, if the lowpass filter function $\ell_{\omega_0}(t)$ in (2.1.12) is used as a window function, then the windowing process (2.2.1) gives rise to an ideal lowpass filter (2.1.10), mapping any finite-energy signal $f(t)$ to a bandlimited signal $f_{\omega_0}(t)$ with bandwidth $\leq \omega_0$.

However, as mentioned earlier, the lowpass filter function $\ell_{\omega_0}(t)$ has a very slow decay and is, therefore, not suitable for the purpose of windowing (in the time domain). To be more precise on the specifications of a window function, we need the terminologies of "center" and "width" of the window function $h(t)$.

Let $h(t)$ be a nontrivial function with sufficiently rapid decay at infinity, so that the following integrals are finite. Then the *center* of the function $h(t)$ is defined by

$$t^* := \frac{\int_{-\infty}^{\infty} t|h(t)|^2 dt}{\int_{-\infty}^{\infty} |h(t)|^2 dt}, \tag{2.2.2}$$

and the *radius* of $h(t)$ is given by

$$\Delta_h := \left\{ \frac{\int_{-\infty}^{\infty} (t - t^*)^2 |h(t)|^2 dt}{\int_{-\infty}^{\infty} |h(t)|^2 dt} \right\}^{1/2}. \tag{2.2.3}$$

Assuming that Δ_h is finite, the width of the window function is set to be

$$2\Delta_h. \tag{2.2.4}$$

In the engineering literature, $2\Delta_h$ is called the RMS duration of $h(t)$. Similarly, if the Fourier transform $\widehat{h}(\omega)$ of $h(t)$ also has sufficiently rapid decay at infinity, so that both of the values

$$\omega^* := \frac{\int_{-\infty}^{\infty} \omega|\widehat{h}(\omega)|^2 d\omega}{\int_{-\infty}^{\infty} |\widehat{h}(\omega)|^2 d\omega} \tag{2.2.5}$$

and

$$\Delta_{\widehat{h}} := \left\{ \frac{\int_{-\infty}^{\infty} (\omega - \omega^*)^2 |\widehat{h}(\omega)|^2 d\omega}{\int_{-\infty}^{\infty} |\widehat{h}(\omega)|^2 d\omega} \right\}^{1/2} \tag{2.2.6}$$

are finite, then the center and width of $\widehat{h}(\omega)$ are given by ω^* and $2\Delta_{\widehat{h}}$, respectively. The value $2\Delta_{\widehat{h}}$ is usually called the RMS bandwidth of the window function $h(t)$.

Now, if $\Delta_{\widehat{h}}$ is finite, we call $\widehat{h}(\omega)$ a frequency window. Similarly, if Δ_h is finite, the function $h(t)$ is called a time window. If both Δ_h and $\Delta_{\widehat{h}}$ are finite, then $h(t)$ provides a time-frequency window.

It is easy to see that the lowpass and bandpass filters $\ell_{\omega_0}(t)$ and $b_{\omega_{n-1}, \omega_n}(t)$ have infinite RMS durations. Hence, although they give rise to "ideal" frequency-domain partitioning, they are not time-frequency windows (since they do not give good time localization).

The *uncertainty principle* says that any time-frequency window $h(t)$ must satisfy the inequality

$$\Delta_h \Delta_{\widehat{h}} \geq \frac{1}{2}, \tag{2.2.7}$$

and that equality in (2.2.7) holds when and only when $h(t)$ is the function

$$h(t) = ce^{jat} e^{-(t-b)^2/4\alpha} \tag{2.2.8}$$

for some constants a, b, c, and α with $\alpha > 0$ and $c \neq 0$. In other words, any time-frequency window cannot have an area smaller than 2, and any *Gaussian function*, as given by (2.2.8), is the only window function with an optimal (or smallest) time-frequency window.

It remains to explain why $h(t)$ is called a time-frequency window. Let us first recall a very useful identity in Fourier transform theory, called the *Parseval identity*:

$$\langle f, g \rangle = \frac{1}{2\pi} \langle \hat{f}, \hat{g} \rangle \tag{2.2.9a}$$

or, explicitly,

$$\int_{-\infty}^{\infty} f(t)\overline{g(t)} \, dt = \frac{1}{2\pi} \int_{-\infty}^{\infty} \hat{f}(\omega)\overline{\hat{g}(\omega)} \, d\omega. \tag{2.2.9b}$$

Returning to the windowing process described by (2.2.1), we see that

$$\int_{-\infty}^{\infty} f(t)\overline{h(t-b)} \, dt = \frac{1}{2\pi} \int_{-\infty}^{\infty} \hat{f}(\omega)e^{jb\omega} \, \overline{\hat{h}(\omega)} \, d\omega. \tag{2.2.10}$$

Observe that the function $h(t-b)$ in the left-hand side localizes the signal $f(t)$ "near $t = b$." More precisely, the time window of the left-hand side in (2.2.10) is given by

$$[b + t^* - \Delta_h, \; b + t^* + \Delta_h], \tag{2.2.11}$$

where t^* is the center of $h(t)$ as defined in (2.2.2). That is, the center of the window function $h(t-b)$ is at $t = b + t^*$ and the radius is given by Δ_h. Similarly, the function $\frac{1}{2\pi}e^{-jb\omega}\hat{h}(\omega)$ in the right-hand side of (2.2.10) localizes the spectrum $\hat{f}(\omega)$ of the signal with frequency window

$$[\omega^* - \Delta_{\hat{h}}, \; \omega^* + \Delta_{\hat{h}}]. \tag{2.2.12}$$

In applications, we only consider real-valued time-window functions. Such functions $h(t)$ clearly satisfy

$$\hat{h}(-\omega) = \overline{\hat{h}(\omega)}, \tag{2.2.13}$$

so that $|\hat{h}(\omega)|$ is an even function. That is, the center ω^* of $\hat{h}(\omega)$ is located at $\omega = 0$, and the frequency window in (2.2.12) becomes

$$[-\Delta_{\hat{h}}, \; \Delta_{\hat{h}}]. \tag{2.2.14}$$

For this reason, we also say that the time-windowing process in (2.2.1) maps a finite-energy signal $f(t)$ to a signal with RMS bandwidth $2\Delta_{\hat{h}}$. Observe that the RMS bandwidth of the ideal lowpass filter with standard bandwidth Ω is given by

$$2\Delta_{\chi_{[-\Omega,\Omega]}} = 2 \left\{ \frac{\int_{-\Omega}^{\Omega} \omega^2 d\omega}{\int_{-\Omega}^{\Omega} d\omega} \right\}^{1/2} \tag{2.2.15}$$

$$= 2 \left\{ \frac{\frac{\Omega^3}{3}}{\Omega} \right\}^{1/2} = \frac{2}{\sqrt{3}}\Omega.$$

Hence, the notion of RMS bandwidth for bandlimited signals is somewhat related to the standard bandwidth. Observe, however, that the definition of the standard bandwidth in (2.1.6) differs from that of the RMS bandwidth in (2.2.14) in that the factor 2 is not used in (2.1.6). The reason that $2\Delta_{\widehat{h}}$ instead of $\Delta_{\widehat{h}}$ is used for RMS bandwidth is that the windowing process introduced in this section will be modified to give bandpass filtering in what is usually called the STFT to be discussed in the next section.

2.3. The short-time Fourier transform.

It is important to point out the difference between the time window (2.2.11) and its corresponding frequency window (2.2.12). While the time window can be slid along the entire time axis by changing the values of b, the frequency window (2.2.12) does not move along the frequency axis at all! It only localizes the frequency range with center at ω^*.

In the following, let us only consider real-valued time-window functions $h(t)$. Hence, from the filtering point of view, the windowing process (2.2.1) maps any finite-energy signal $f(t)$ to an RMS bandlimited signal

$$f_0(t) = \int_{-\infty}^{\infty} f(x)h(x-t)dx \qquad (2.3.1)$$

that lives in the (RMS) frequency band

$$[-\Delta_{\widehat{h}}, \ \Delta_{\widehat{h}}]. \qquad (2.3.2)$$

This may be considered a generalization of the (ideal) lowpass filtering with filter function $\ell_{\omega_0}(t)$. However, for the window function $h(t)$ to be a legitimate lowpass filter, we really should require

$$|\widehat{h}(0)| = 1, \qquad (2.3.3)$$

so that at least the DC content of the signal (i.e., at the zero frequency) is passed. One of the lowpass filter design criteria is to require $|\widehat{h}(\omega)|$ to be somewhat *flat* at $\omega = 0$; namely,

$$\left[\frac{d^i}{d\omega^i} |\widehat{h}(\omega)| \right]_{\omega=0} = 0, \qquad i = 1, \ldots, m \qquad (2.3.4)$$

for some positive integer m. That is, $h(t)$ is a "Butterworth filter" of order m.

Let us now find a means to slide the frequency window (2.2.14) or, more generally, (2.2.12). Recall that a translation in the time domain corresponds to a phase shift in the frequency domain. Similarly, a phase shift in the time domain also corresponds to a translation in the frequency domain. Hence, if we wish to slide the frequency window in (2.2.14) along the frequency axis, all we have to do is to incorporate a phase shift in the windowing process described by (2.2.1). This leads to the following notion of STFT.

From now on, we will use the notation $\phi(t)$ for a real-valued lowpass window function and require that $\phi(t)$ satisfies $\widehat{\phi}(0) = 1$.

Definition 2.1. *Let $\phi(t)$ be a real-valued function with $\phi(t)$, $|t|^{1/2}\phi(t)$, and $t\phi(t)$ in L^2 such that $\hat{\phi}(0) = 1$. Then the STFT with window function $\phi(t)$ is defined by*

$$(G_\phi f)(b, \xi) := \int_{-\infty}^{\infty} f(t)e^{-j\xi t}\phi(t - b)dt, \quad f \in L^2. \tag{2.3.5}$$

The STFT was first introduced by D. Gábor who used the Gaussian function

$$g_\alpha(t) = \frac{1}{2\sqrt{\pi\alpha}}e^{-t^2/4\alpha}, \tag{2.3.6}$$

with $\alpha > 0$, as the lowpass window $\phi(t)$. Hence, (2.3.5) is also called the Gabor transform in the literature.

With the additional phase shift in (2.3.5), an application of the Parseval identity (2.2.9b) now gives

$$(G_\phi f)(b, \xi) = \int_{-\infty}^{\infty} f(t)e^{-j\xi t}\phi(t - b)dt \tag{2.3.7}$$

$$= \frac{e^{-j\xi b}}{2\pi} \int_{-\infty}^{\infty} \hat{f}(\omega)e^{jb\omega} \overline{\hat{\phi}(\omega - \xi)}\, d\omega.$$

The improvement in the transform G_ϕ in (2.3.5) over the simple sliding time-windowing process in (2.2.1) is that the frequency-window function $\hat{\phi}(\omega)$ is now able to slide along the entire frequency axis, so that the lowpass window function $\phi(t)$ can be used for bandpass filtering. More precisely, while the modulated time-window function

$$e^{j\xi t}\phi(t - b) \tag{2.3.8}$$

in (2.3.7) as a function of t localizes the signal $f(t)$ "near $t = b$," the corresponding modulated frequency window function

$$\frac{e^{j\xi b}}{2\pi}e^{-jb\omega} \hat{\phi}(\omega - \xi) \tag{2.3.9}$$

in (2.3.7) as a function of ω localizes the spectrum (or Fourier transform) $\hat{f}(\omega)$ of the signal around $\omega = \xi$, with frequency window

$$[\xi - \Delta_{\hat{\phi}},\ \xi + \Delta_{\hat{\phi}}]. \tag{2.3.10}$$

If the center t^* of $\phi(t)$ is located at the origin, then the STFT in (2.3.5) localizes the analog signal around $t = b$, with time window

$$[b - \Delta_\phi,\ b + \Delta_\phi]. \tag{2.3.11}$$

Otherwise, a shift of $\phi(t)$ by t^* to the left gives the same result. More precisely, the *centered STFT* defined by

$$(G_\phi^c f)(b, \xi) := \int_{-\infty}^{\infty} f(t)e^{-j\xi t}\phi(t + t^* - b)dt \tag{2.3.12}$$

has time window given by (2.3.11) and frequency window again by (2.3.10). The only change in the modulated frequency-window function (2.3.9) is to replace b by $b - t^*$.

In Figure 2.5, we show the time-frequency window corresponding to the window function $\phi(t)$ at various time-frequency locations (b, ξ). Observe that the area of the window is always given by $4\Delta_\phi\Delta_{\hat{\phi}}$.

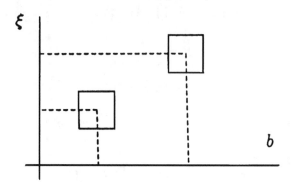

FIG. 2.5.   *Time-frequency window of the STFT.*

2.4. The integral wavelet transform.

We emphasize that the STFT G_ϕ, where ϕ is a lowpass window function, even has the capability of bandpass filtering when the window is allowed to slide along the frequency axis. One main drawback of this window is that its size is rigid. This means that the width of the time window remains the same for studying low-frequency phenomena or detecting high-frequency changes. The analog signal in Figure 2.6 has low-frequency content on the left and high-frequency content on the right of the time axis. Hence, to study this so-called chirp signal accurately, it is desirable for the time window to narrow as it slides from left to right, as shown in Figure 2.7.

In the following, we will introduce the notion of IWT (also called CWT) that provides a flexible time-frequency window. By mapping the *scale param-eter* to the frequency variable, this window automatically widens to provide a better understanding of low-frequency environments and narrows to give a more precise detection of high-frequency changes.

Recall that in formulating the STFT, a phase shift was applied to the sliding time-windowing process (or inner product) in (2.2.1) to enable the frequency window to slide along the frequency axis as shown in (2.3.7). In formulating the IWT, a scale parameter is introduced to adjust the width of the sliding time-windowing process instead. It will be seen that this parameter is also used to slide the frequency window along the frequency axis. More precisely, let us consider the integral transform

$$\int_{-\infty}^{\infty} f(t)\overline{\psi\left(\frac{t-b}{a}\right)}\, dt, \qquad (2.4.1)$$

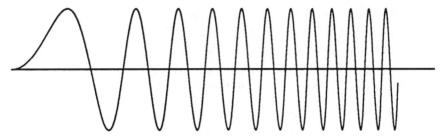

FIG. 2.6. *Chirp signal (with frequency increasing with time).*

FIG. 2.7. *Flexible time window to zoom in.*

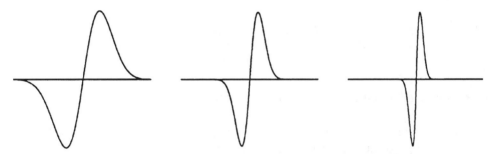

FIG. 2.8. *The time window narrows with decreasing values of the scale a.*

where a is the scale parameter. In Figure 2.8, we demonstrate how the width of the window function $\psi(\frac{t-b}{a})$ is adjusted by changing the values of the scale $a > 0$.

Observe that we have used $\psi(t)$ instead of the notation $h(t)$ in (2.4.1) in order to emphasize the necessary requirements on this window function which will be discussed later. For this windowing process to be effective, we will again require $\psi(t)$ to be real valued, but instead of using a lowpass window function, $\psi(t)$ will be assumed to be bandpass in the sense that

$$\widehat{\psi}(0) = 0 \qquad (2.4.2)$$

(that is, at least the zero frequency is stopped). In other words, the window function $\psi(t)$ has zero mean, and that is why it is called a wavelet, meaning that its graph is a small wave.

This requirement is needed for perfect recovery of the signal from its IWT. As we have already pointed out in (2.2.13), the importance for $\psi(t)$ to be a real-valued function is that its Fourier transform $\widehat{\psi}(\omega)$ satisfies

$$\widehat{\psi}(-\omega) = \overline{\widehat{\psi}(\omega)}, \qquad (2.4.3)$$

so that $|\widehat{\psi}(\omega)|$ is an even function. Now, since we are only interested in non-negative values of the frequency variable ω, and since $\psi(t)$ is a bandpass filter, we only consider $\widehat{\psi}(\omega)$ as a frequency window on the frequency domain $[0, \infty)$. This consideration requires the following modification of the notion of centers and widths of the frequency-window function $\widehat{\psi}(\omega)$.

Definition 2.2. *Let $\psi(t)$ be a real-valued function, such that $\psi(t)$, $|t|^{1/2}\psi(t)$, and $t\psi(t)$ are in L^2 and $\widehat{\psi}(0) = 0$. Then the (one-sided) center of $\widehat{\psi}(\omega)$, considered as a function on $[0, \infty)$, is defined by*

$$\omega_+^* := \frac{\int_0^\infty \omega |\widehat{\psi}(\omega)|^2 d\omega}{\int_0^\infty |\widehat{\psi}(\omega)|^2 d\omega}, \tag{2.4.4}$$

and the (one-sided) radius of $\widehat{\psi}(\omega)$ is defined by

$$\Delta_{\widehat{\psi}}^+ := \left(\frac{\int_0^\infty (\omega - \omega_+^*)^2 |\widehat{\psi}(\omega)|^2 d\omega}{\int_0^\infty |\widehat{\psi}(\omega)|^2 d\omega} \right)^{1/2}. \tag{2.4.5}$$

The one-sided width of the frequency window $\widehat{\psi}(\omega)$ is then given by $2\Delta_{\widehat{\psi}}^+$.

For a real-valued window function $\psi(t)$, by introducing a normalization factor $a^{-1/2}$, the integral transform (2.4.1) becomes the IWT or CWT defined as follows.

Definition 2.3. *Let $\psi(t)$ be a real-valued function that satisfies the properties stated in Definition 2.2. Then the IWT or CWT, with window function $\psi(t)$, is defined by*

$$(W_\psi f)(b, a) := \frac{1}{\sqrt{a}} \int_{-\infty}^\infty f(t)\psi\left(\frac{t - b}{a}\right) dt, \qquad f \in L^2, \tag{2.4.6}$$

where $a > 0$. The bandpass window function $\psi(t)$ in (2.4.6) is called the analyzing wavelet of the IWT.

First, it is important to ensure that the IWT W_ψ gives rise to frequency localization by mapping scale to frequency and that the corresponding frequency window also "slides" along the (positive) frequency axis.

In the following, to facilitate our presentation, we will only consider finite-energy signals $f(t)$ that are real valued, and the notation for the class of such functions will be

$$L_R^2 = \{f(t) \in L^2 \colon f(t) \text{ real}\}. \tag{2.4.7}$$

Now, since $\psi(t)$ is also real, we have

$$\widehat{f}(-\omega)e^{jb\omega}\overline{\widehat{\psi}(-a\omega)} = \overline{\widehat{f}(\omega)e^{-jb\omega}\overline{\widehat{\psi}(a\omega)}}, \qquad f \in L_R^2. \tag{2.4.8}$$

In view of this observation, we can now apply the Parseval identity (2.2.9b) to obtain

$$(W_\psi f)(b, a) = \frac{1}{\sqrt{a}} \int_{-\infty}^{\infty} f(t) \psi\left(\frac{t - b}{a}\right) dt \qquad (2.4.9)$$

$$= \frac{\sqrt{a}}{2\pi} \int_{-\infty}^{\infty} \hat{f}(\omega) e^{-jb\omega} \overline{\hat{\psi}(a\omega)} \, d\omega$$

$$= \frac{\sqrt{a}}{\pi} \, \mathrm{Re} \int_{0}^{\infty} \hat{f}(\omega) e^{-jb\omega} \overline{\hat{\psi}(a\omega)} \, d\omega$$

$$= \frac{\sqrt{a}}{\pi} \, \mathrm{Re} \int_{0}^{\infty} \hat{f}(\omega) e^{-jb\omega} \overline{\eta\left(a\left(\omega - \frac{\omega_+^*}{a}\right)\right)} \, d\omega,$$

for all $f \in L_R^2$, where we have introduced the notation

$$\begin{cases} \eta(\omega) := \hat{\psi}(\omega + \omega_+^*), \quad \text{with} \\ \omega_+^* \text{ denoting the (one-sided) center of } \hat{\psi}(\omega) \text{ on } [0, \infty). \end{cases} \qquad (2.4.10)$$

Note that the center of the frequency-window function $\eta(\omega)$ is located at the origin.

To map the scale parameter a to the (positive) frequency axis, we may consider

$$a \longmapsto \xi := \frac{c}{a} \quad \text{for some} \quad c > 0. \qquad (2.4.11)$$

The constant c here is called a *calibration constant* in frequency units (such as Hz). With the change of variables given in (2.4.11), we see that the IWT, which localizes a signal $f(t)$ in the time domain, also localizes its spectrum $\hat{f}(\omega)$ in the frequency domain simultaneously; namely,

$$(W_\psi f)(b, \xi) = \frac{\sqrt{a}}{\pi} \mathrm{Re} \int_{0}^{\infty} \hat{f}(\omega) e^{-jb\omega} \overline{\eta\left(a\left(\omega - \frac{\omega_+^*}{c}\xi\right)\right)} \, d\omega. \qquad (2.4.12)$$

Observe that the frequency window of the window function $\eta\left(a\left(\omega - \frac{\omega_+^*}{c}\xi\right)\right)$ is given by

$$\left[\frac{\omega_+^*}{c}\xi - \frac{1}{a}\Delta_{\hat{\psi}}^+, \; \frac{\omega_+^*}{c}\xi + \frac{1}{a}\Delta_{\hat{\psi}}^+\right] = \left[\frac{1}{a}\left(\omega_+^* - \Delta_{\hat{\psi}}^+\right), \; \frac{1}{a}\left(\omega_+^* + \Delta_{\hat{\psi}}^+\right)\right]. \qquad (2.4.13)$$

The width of this window, which is $\frac{2}{a}\Delta_{\hat{\psi}}^+$, is independent of the calibration constant c. Also observe that by choosing the wavelet $\psi(t)$ so that

$$\omega_+^* > \Delta_{\hat{\psi}}^+ \qquad (2.4.14)$$

is satisfied, this window moves up the frequency axis with increasing values of ξ as defined by (2.4.11) or decreasing values of the scale a, but the window itself widens.

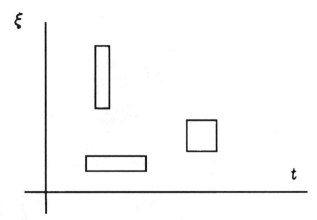

FIG. 2.9.   *Time-frequency window of the IWT (or CWT).*

Although it is very tempting to set $c = \omega_+^*$ to simplify the notations in (2.4.12)–(2.4.13), the choice of the calibration constant c should really depend on the applications. One method to determine this constant is to take the IWT of a (truncated) pure sinusoidal function with a single (but known) frequency and to match this (frequency) value with the scale axis.

On the other hand, the time-window function $\frac{1}{\sqrt{a}}\psi\left(\frac{t-b}{a}\right)$ in (2.4.6) is given by

$$[b + at^* - a\Delta_\psi, \ b + at^* + a\Delta_\psi], \qquad (2.4.15)$$

where t^* is the center of $\psi(t)$ as defined in (2.2.2). The width of this time window is $2a\Delta_\psi$ which decreases at high frequencies ξ (or small values of $a > 0$) and increases at low frequencies ξ. It follows from (2.4.13) and (2.4.15) that the time-frequency window of the IWT (with a real-valued bandpass window function ψ as the analyzing wavelet) is given by

$$[b + at^* - a\Delta_\psi, \ b + at^* + a\Delta_\psi] \times \left[\frac{\omega_+^*}{c}\xi - \frac{1}{a}\Delta_{\hat{\psi}}^+, \ \frac{\omega_+^*}{c}\xi + \frac{1}{a}\Delta_{\hat{\psi}}^+\right]. \qquad (2.4.16)$$

We remark that although the area of the window (2.4.16) is a constant, given by

$$(2a\Delta_\psi)\left(\frac{2}{a}\Delta_{\hat{\psi}}^+\right) = 4\Delta_\psi\Delta_{\hat{\psi}}^+, \qquad (2.4.17)$$

this window automatically changes in size to adapt to the frequencies of the signal under investigation. The time-frequency plot of the window (2.4.16) is shown in Figure 2.9.

As in the discussion of time localization by the STFT in (2.3.12), we may also center the time window of the IWT by introducing the *centered* IWT, W_ψ^c, defined by

$$(W_\psi^c f)(b, a) = \frac{1}{\sqrt{a}} \int_{-\infty}^{\infty} f(t)\psi\left(\frac{t-b}{a} + t^*\right)dt, \qquad f \in L^2, \qquad (2.4.18)$$

where t^* is the center of the window function $\psi(t)$. The time-frequency window of this centered IWT, W_ψ^c, is then given by

$$\left[b - a\Delta_\psi, \; b + a\Delta_\psi\right] \times \left[\frac{\omega_+^*}{c}\xi - \frac{1}{a}\Delta_{\widehat{\psi}}^+, \; \frac{\omega_+^*}{c}\xi + \frac{1}{a}\Delta_{\widehat{\psi}}^+\right]. \tag{2.4.19}$$

With the change of $\Delta_{\widehat{\psi}}$ in (2.2.6) (for $\widehat{h} = \widehat{\psi}$) to $\Delta_{\widehat{\psi}}^+$ in (2.4.5), the uncertainty principle (2.2.7) stated in section 2.2 needs certain modification. It turns out, however, that the lower bound is still $\frac{1}{2}$. More precisely, the uncertainty principle for bandpass filtering is

$$\Delta_\psi \Delta_{\widehat{\psi}}^+ > \frac{1}{2}. \tag{2.4.20}$$

We remark that $\frac{1}{2}$ is the largest lower bound, but this value cannot be attained by any $\psi(t)$. A study of uncertainty bounds will be addressed in Chapter 5.

2.5. Modeling the cochlea.

To demonstrate the significance of the IWT in signal analysis, let us study the following mathematical model of the spiral-like organ, called the cochlea, located behind the eardrum of the human ear (see Figure 2.10).

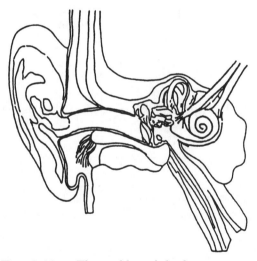

FIG. 2.10. *The cochlea of the human ear.*

Let x denote the location of a certain sensory cell in the corti of the cochlea. Hence, if we consider the audio reception $g_x(t)$ (output) at x as the convolution of an acoustic signal $f(t)$ (input) with respect to some filter function (which depends on the location x), we have

$$\widehat{g}_x(\omega) = H_x(\omega)\widehat{f}(\omega), \tag{2.5.1}$$

where $H_x(\omega)$ is the transfer function of this linear filter. Let us assume that the spiral geometry of the cochlea induces a logarithmic delay in frequency.

(This assumption happens to agree with the empirical result of Zweig.) Then the transfer function $H_x(\omega)$ at x is given by

$$H_x(\omega) = G(x - \ln \omega) \qquad (2.5.2)$$

for some function $G(x)$. Hence, by considering

$$a := e^{-x}, \qquad (2.5.3)$$

or $x = \ln(\frac{1}{a})$, and by introducing some real-valued function $\psi(t)$ through its Fourier transform

$$\widehat{\psi}(\omega) := \overline{G\left(\ln\left(\frac{1}{\omega}\right)\right)}, \qquad (2.5.4)$$

we see that the input–output relation in (2.5.1) is given by

$$\widehat{g}_x(\omega) = \overline{\widehat{\psi}(a\omega)}\, \widehat{f}(\omega). \qquad (2.5.5)$$

Hence, taking the inverse Fourier transform, we arrive at

$$g_x(t) = \frac{1}{2\pi} \int_{-\infty}^{\infty} e^{j\omega t}\, \overline{\widehat{\psi}(a\omega)}\, \widehat{f}(\omega) d\omega \qquad (2.5.6)$$

$$= \frac{1}{2\pi} \int_{-\infty}^{\infty} \widehat{f}(\omega)\, \overline{\widehat{\psi}(a\omega) e^{-j\omega t}}\, d\omega$$

$$= \frac{1}{a} \int_{-\infty}^{\infty} f(s) \psi\left(\frac{s-t}{a}\right) ds,$$

where the Parseval identity (2.2.9b) is used. That is, the reception at the sensory cell located at x is given by

$$g_x(t) = e^{-x/2}(W_\psi f)(t,\ e^{-x}). \qquad (2.5.7)$$

To ensure that $W_\psi f$ in (2.5.7) is the IWT of $f(t)$, we need ψ to have zero mean. This follows from the assumption that there is no reception at $x = \infty$; namely,

$$\int_{-\infty}^{\infty} \psi(t) dt = \widehat{\psi}(0) = \overline{G(\infty)} = \overline{H_\infty(\omega)} = 0. \qquad (2.5.8)$$

The problem that remains is to construct the analyzing wavelet $\psi(t)$.

CHAPTER **3**

Multiresolution Analysis

Both the short-time Fourier transform (STFT) and the integral wavelet transform (IWT), also known as continuous wavelet transform (CWT), are very useful time-frequency localization tools, provided that the lowpass window function $\phi(t)$ for the STFT and the bandpass window function $\psi(t)$ for the IWT are suitably chosen and well localized, with not-so-large values of

$$\Delta_\phi \Delta_{\hat{\phi}} < \infty \quad \text{and} \quad \Delta_\psi \Delta_{\hat{\psi}}^{\pm} < \infty.$$

In applying the IWT with $\psi(t)$ as the analyzing wavelet to signal analysis, we first map the finite-energy analog signal into some signal space V_ϕ which is determined by a lowpass window function $\phi(t)$ with finite RMS bandwidth. For computational and implementational efficiency, this lowpass filter $\phi(t)$ must satisfy a *two-scale relation*. That is, $\phi(t)$ must be a *scaling function*. Such a scaling function generates a multiresolution analysis (MRA) of the space L^2 of finite-energy signals.

In the first section of this chapter we describe how a finite-energy signal is represented as a signal in V_ϕ in terms of superpositions of translates of ϕ on a uniform sample set. The concept of stability of such signal representations will also be discussed in this section. A study of "scaling functions" to take care of nonuniform sample sets will be addressed in Chapter 6. Two very simple examples will be given in section 3.2: the first by applying the sampling theorem and the second by considering piecewise constant interpolation. These two examples will be used to motivate the notion of MRA in section 3.3, and the example of piecewise constant representation will also be generalized to the representation by cardinal splines in section 3.4, demonstrating the effectiveness of the MRA approach in signal representations and signal analysis.

3.1. Signal spaces with finite RMS bandwidth.

It is well known that every bandlimited signal can be perfectly recovered from its discrete samples, provided that the sampling period is sufficiently small. This is a consequence of the *sampling theorem*, which can be stated as follows.

Theorem 3.1. *Let $f(t) \in L^2$ be a bandlimited signal with bandwidth $\leq \Omega$ where $\Omega > 0$, and let*

$$0 < a \leq \frac{\pi}{\Omega}. \tag{3.1.1}$$

Then $f(t)$ is a continuous function and can be perfectly recovered from its digital samples

$$f(ka), \qquad k = 0, \pm 1, \ldots, \tag{3.1.2}$$

by using the formula

$$f(t) = \sum_{k=-\infty}^{\infty} f(ka)\phi_{s;a}(t - ka), \tag{3.1.3}$$

where

$$\phi_{S;a}(t) := \phi_S\left(\frac{t}{a}\right), \tag{3.1.4}$$

and $\phi_S(t)$ denotes the Shannon sampling function

$$\phi_S(t) := \frac{\sin \pi t}{\pi t}. \tag{3.1.5}$$

Recall that the Shannon sampling function $\phi_S(t)$ is nothing but the lowpass filter function $\ell_{\omega_0}(t)$ in (2.1.12) with $\omega_0 = \pi$, which has an ideal filter characteristic given by

$$\hat{\phi}_S(\omega) = \chi_{[-\pi, \pi]}(\omega). \tag{3.1.6}$$

See (2.1.7)–(2.1.12). However, since $\phi_S(t)$ has infinite RMS duration, it is not suitable for time localization. For this reason, we will replace $\phi_S(t)$ by a lowpass time-frequency window function $\phi(t)$ with finite value of $\Delta_\phi \Delta_{\hat{\phi}}$ but retain the structure of the signal representation in the sampling theorem. More precisely, we will consider the *signal space* of functions of the form

$$\sum_k c_k \phi(t - ka), \tag{3.1.7}$$

where $a > 0$ is a preassigned *sampling period*, and c_k's are arbitrary constants so chosen that the representation (3.1.7) remains a finite-energy signal. In mathematical language, this signal space is called the L^2-closure of all finite sums of the form (3.1.7), and the notation

$$V_{\phi,a} := \text{clos}_{L^2}\langle \phi(t - ka) \colon k \in \mathbb{Z}\rangle \tag{3.1.8}$$

will be used (recall (1.2.2)).

One of the most important engineering issues is to study the *stability* of the signal representations as given in (3.1.7). Here, stability means that a small disturbance of the coefficient sequence $\{c_k\}$ in (3.1.7) should only result in a small disturbance of the signal it represents and vice versa. Let us use the energy (or L^2-) norm in (2.1.3) to measure the change in signal content and the discrete energy (or ℓ^2 sequence) norm

$$\|\{c_k\}\|_{\ell^2} := \left\{ \sum_k |c_k|^2 \right\}^{1/2} \tag{3.1.9}$$

to measure the square root of the total energy of the corresponding coefficient sequence. Hence, if we consider two signal representations

$$\begin{cases} f_1(t) = \sum_k a_k \phi(t - ka), \\ f_2(t) = \sum_k b_k \phi(t - ka), \end{cases}$$

then we are interested in comparing the errors

$$\|f_1 - f_2\| \quad \text{and} \quad \|\{a_k - b_k\}\|_{\ell^2}.$$

By setting

$$\begin{cases} f(t) = f_1(t) - f_2(t), \\ c_k = a_k - b_k \end{cases}$$

so that

$$f(t) = \sum_k c_k \phi(t - ka), \tag{3.1.10}$$

the stability consideration is to compare the two quantities

$$\left\| \sum_k c_k \phi(t - ka) \right\| \quad \text{and} \quad \|\{c_k\}\|_{\ell^2}. \tag{3.1.11}$$

This leads to the following notion of *Riesz (or stable) basis*. For convenience, we will consider $a = 1$.

Definition 3.1. *A function $\phi(t) \in L^2$ is said to generate a Riesz basis (or stable basis)*

$$\{\phi(t - k) : k = 0, \pm 1, \ldots\} \tag{3.1.12}$$

of the space

$$V_\phi := \text{clos}_{L^2} \langle \phi(t - k) : k \in \mathbb{Z} \rangle, \tag{3.1.13}$$

if there exist two positive constants A and B, with $0 < A \le B < \infty$, called Riesz bounds such that

$$A\|\{c_k\}\|_{\ell^2}^2 \le \left\| \sum_k c_k \phi(t - k) \right\|^2 \le B\|\{c_k\}\|_{\ell^2}^2 \tag{3.1.14}$$

for all sequences $\{c_k\} \in \ell^2$ *(that is, sequences with finite energy* $\|\{c_k\}\|_{\ell^2}^2$*). If* $\phi(t)$ *satisfies* (3.1.14)*, it is called a stable function.*

Observe that the inequalities in (3.1.14) guarantee that the two energy quantities in (3.1.11), with $a = 1$, are simultaneously small.

Example 3.1. Recall that the Gaussian function

$$h_\alpha(t) := \frac{1}{2\alpha\sqrt{\pi}} e^{-t^2/4\alpha^2} \qquad (3.1.15)$$

is a lowpass window function that provides optimal time-frequency localization as governed by the uncertainty principle. (See (2.2.7) and (2.2.8) with $a = b = 0$ and $c = 1/(2\alpha\sqrt{\pi})$.) It is not difficult to verify that $h_\alpha(t)$ generates a Riesz basis of V_{h_α} with Riesz bounds A and B satisfying

$$\frac{1}{\pi} \int_\pi^\infty e^{-2\alpha^2 x^2} dx \le A < B \le 1 + \frac{1}{\sqrt{2\pi}\,\alpha}. \qquad (3.1.16)$$

Another important issue is to devise a mapping from the finite-energy signal space L^2 into the RMS bandlimited space $V_{\phi,a}$ (such as $V_\phi := V_{\phi,1}$). For the sampling function $\phi_{S;a}(t)$ in (3.1.4)–(3.1.5), the Shannon sampling operator

$$(V_a^S f)(t) := \sum_k f(ka)\phi_{S;a}(t - ka) = \sum_k f(ka)\phi_S\left(\frac{t}{a} - k\right), \qquad (3.1.17)$$

which maps $f(t) \in L^2$ into $V_{\phi_{S;a},a}$, makes sense only if $f(t)$ is a continuous function. Also, the mapping V_a^S provides a fairly accurate signal representation only if the finite-energy signal $f(t)$ is bandlimited. In fact, if the bandwidth of $f(t)$ is Ω, then according to the sampling theorem (see Theorem 3.1) we have

$$(V_a^S f)(t) \equiv f(t) \qquad (3.1.18)$$

for any sampling period $a \le \frac{\pi}{\Omega}$.

For an arbitrary lowpass filter function $\phi(t)$, however, the corresponding *sampling operator*

$$(V_a f)(t) := \sum_k f(ka)\phi(t - ka) \qquad (3.1.19)$$

usually does not give a good representation of $f(t) \in L^2$, even if the sampling period a is small. (See Example 6.2 in Chapter 6.) Here, observe that $f(t)$ must be a continuous function for the sample sequence $\{f(ka)\}$ to be meaningful.

So, what mapping gives a good signal representation? The most natural mapping is probably the so-called orthogonal projection (or L^2-projection). Let us denote this map to be L_a. Then the orthogonal projection $(L_a f)(t)$ of $f(t)$ to the space $V_{\phi,a}$ is the function in $V_{\phi,a}$ which is closest to $f(t)$, using the measurement $\|\ \|$. In other words, the error function $f(t) - (L_a f)(t)$ is

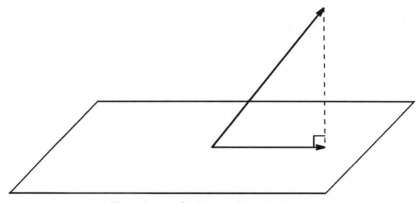

FIG. 3.1. *Orthogonal projection.*

orthogonal to all of $V_{\phi,a}$ (see Figure 3.1). Hence, the mathematical formulation of the L^2-projection mapping L_a is given by

$$\begin{cases} (L_a f)(t) = \sum_k c_k \phi(t - ka) & \text{such that} \\ \int_{-\infty}^{\infty} \{f(t) - (L_a f)(t)\} \phi(t - ka) dt = 0, & k = 0, \pm 1, \ldots . \end{cases} \qquad (3.1.20)$$

That is, the coefficient sequence $\{c_k\}$ that defines the L^2-projection mapping L_a is uniquely determined by the orthogonality condition in (3.1.20). To compute $\{c_k\}$, the system of linear equations

$$M\mathbf{c} = \mathbf{b}, \qquad (3.1.21)$$

where

$$\begin{cases} M = \left[\int_{-\infty}^{\infty} \phi(t - ka)\phi(t - \ell a) dt \right]_{-\infty < k, \ell < \infty}, \\ \\ \mathbf{c} = \begin{bmatrix} \vdots \\ c_k \\ c_{k+1} \\ \vdots \end{bmatrix}, \quad \text{and} \quad \mathbf{b} = \begin{bmatrix} \vdots \\ \int_{-\infty}^{\infty} f(t)\phi(t - ka) dt \\ \vdots \end{bmatrix}, \end{cases} \qquad (3.1.22)$$

has to be solved.

Here, M is a bi-infinite Toeplitz matrix and must be inverted. In the special case when M is the identity matrix, then the solution is simply $\mathbf{c} = \mathbf{b}$ or

$$c_k = \int_{-\infty}^{\infty} f(t)\phi(t - ka) dt, \qquad k = 0, \pm 1, \ldots . \qquad (3.1.23)$$

Observe that M is the identity matrix, if and only if $\{\phi(t - ka)\}$ is an orthonormal family, defined as follows.

Definition 3.2. *A function $\phi(t) \in L^2$ is said to generate an orthonormal family on the sample set $\{ka : k = 0, \pm 1, \ldots\}$ if $\{\phi(t - ka)\}$ satisfies*

$$\int_{-\infty}^{\infty} \phi(t - ka)\phi(t - \ell a)dt = \delta_{k,\ell}, \qquad k, \ell \in \mathbb{Z}. \qquad (3.1.24)$$

Here and throughout, $\delta_{k,\ell}$ denotes the Kronecker delta symbol

$$\delta_{k,\ell} = \begin{cases} 1 & \text{if } k = \ell, \\ 0 & \text{otherwise.} \end{cases} \qquad (3.1.25)$$

If (3.1.24) is satisfied for $a = 1$, then $\phi(t)$ is said to be an orthonormal function.

It is important to observe that *any orthonormal basis is a Riesz (or stable) basis* with Riesz bounds $A = B = 1$.

We will give two examples of orthonormal functions in the next section. Before doing so, let us first discuss the relationship between the sampling operator V_a in (3.1.19), using only the digital samples as coefficients, and the L^2-projection L_a as defined in (3.1.20) for any orthonormal lowpass window function $\phi(t)$. This relationship can be generalized to a nonorthonormal $\phi(t)$ by introducing the *dual* $\tilde{\phi}(t)$ of $\phi(t)$, and this generalization will be discussed in section 5.3 of Chapter 5.

Let $f(t)$ be bandlimited with bandwidth Ω and assume that the lowpass window function $\phi(t)$ generates an orthonormal family on the sample set $\{ka : k = 0, \pm 1, \ldots\}$ with $a > 0$. Then by (3.1.20) and applying the Parseval identity (2.2.9b), we have

$$(L_a f)(t) = \sum_k \left\{ \int_{-\infty}^{\infty} f(t)\phi(x - ka)dt \right\} \phi(t - ka) \qquad (3.1.26)$$

$$= \sum_k \left\{ \frac{1}{2\pi} \int_{-\Omega}^{\Omega} \hat{f}(\omega)\overline{\hat{\phi}(\omega)}e^{jka\omega}d\omega \right\} \phi(t - ka).$$

Hence, if the lowpass filter characteristic of $\phi(t)$ is *almost ideal* in the frequency interval $[-\Omega, \Omega]$, in the sense that

$$\hat{\phi}(\omega) \doteq 1, \quad |\omega| < \Omega, \qquad (3.1.27)$$

then since $\hat{f}(\omega) = 0$ for $|\omega| \geq \Omega$, we have

$$\frac{1}{2\pi} \int_{-\Omega}^{\Omega} \hat{f}(\omega)\overline{\hat{\phi}(\omega)}e^{jka\omega}d\omega \doteq \frac{1}{2\pi} \int_{-\Omega}^{\Omega} \hat{f}(\omega)e^{jka\omega}d\omega \qquad (3.1.28)$$

$$= \frac{1}{2\pi} \int_{-\infty}^{\infty} \hat{f}(\omega)e^{jka\omega}d\omega = f(ka).$$

That is, the expression (3.1.26) becomes

$$(L_a f)(t) \doteq \sum_k f(ka)\phi(t - ka) = (V_a f)(t). \qquad (3.1.29)$$

So, when a high-order Butterworth orthonormal lowpass filter function $\phi(t)$, with $\widehat{\phi}(0) = 1$, is used, we have (3.1.27) for sufficiently small $\Omega > 0$. Hence, the sampling operator provides a good approximation of the L^2-projection L_a for all bandlimited analog signals $f(t)$ with sufficiently small bandwidth.

3.2. Two simple mathematical representations.

Observe that in the sampling theorem any scale a that satisfies (3.1.1) can be used to recover all bandlimited signals $f(t)$ with bandwidth Ω by using the formula (3.1.3). Of course this formula is not valid for signals with larger bandwidths, and certainly one cannot expect to have (3.1.3) if an arbitrary lowpass window function $\phi(t)$ is used. For this and other reasons, it is important to introduce an efficient computational scheme to relate the formulas given by the right-hand side of (3.1.3) for different values of the scale a when $\phi_S(t)$ is replaced by another lowpass filter function $\phi(t)$. That is, we need to find a relation, if any, between

$$\phi\left(\frac{t}{a_1}\right) \quad \text{and} \quad \phi\left(\frac{t}{a_2}\right)$$

for two different values a_1 and a_2 of the scale $a > 0$. In this monograph, we will mainly consider scales that are integer powers of 2, and the relation we look for is the following.

$\phi\left(\frac{t}{2}\right)$ is expressed as superpositions of integer translates of $\phi(t)$; namely,

$$\phi\left(\frac{t}{2}\right) = \sum_k p_k \phi(t - k) \quad \text{all } t \tag{3.2.1}$$

for some sequence $\{p_k\}$ such that the expression on the right-hand side of (3.2.1) is in L^2.

Definition 3.3. *If a function $\phi(t)$ in L^2 satisfies (3.2.1) and is stable, then $\phi(t)$ is called a scaling function and the sequence $\{p_k\}$ in (3.2.1) is called its corresponding two-scale sequence. The identity (3.2.1) itself is called the two-scale relation of $\phi(t)$.*

By applying the Fourier transform to both sides of (3.2.1), we see that the two-scale relation (3.2.1) is equivalent to

$$\widehat{\phi}(\omega) = P(e^{-j\omega/2})\widehat{\phi}\left(\frac{\omega}{2}\right) \quad \text{all } \omega \tag{3.2.2}$$

where

$$P(z) := \frac{1}{2}\sum_k p_k z^k \tag{3.2.3}$$

is called the *two-scale symbol* of the sequence $\{p_k\}$.

Example 3.2. The Shannon sampling function $\phi_S(t)$ defined in (3.1.5) is a scaling function that is orthonormal in the sense of (3.1.24) for $a = 1$.

To verify that $\phi_S(t)$ is a scaling function, we recall that its Fourier transform $\hat{\phi}_S(\omega)$ is the ideal lowpass filter characteristic $\chi_{[-\pi,\pi]}(\omega)$ (see (3.1.6)). Hence, we have

$$\hat{\phi}_S\left(\frac{\omega}{2}\right) = \chi_{[-2\pi,2\pi]}(\omega).$$

Now, the Fourier series $P(e^{-j\omega/2})$ that satisfies (3.2.2) for $\phi(t) = \phi_S(t)$ is the 4π-periodic extension of $\chi_{[-\pi,\pi]}(\omega)$. A simple calculation yields

$$p_k = \begin{cases} 1 & \text{for} \quad k = 0, \\[2mm] (-1)^{\frac{k-1}{2}}\frac{2}{k\pi} & \text{for} \quad \text{odd } k, \\[2mm] 0 & \text{otherwise.} \end{cases}$$

That is, the two-scale relation of the Shannon sampling function is given by

$$\phi_S(t) = \phi_S(2t) + \sum_{k=-\infty}^{\infty} (-1)^k \frac{2}{(2k+1)\pi}\phi_S(2t - 2k - 1). \tag{3.2.4}$$

To show that $\phi_S(t)$ generates an orthonormal basis of V_{ϕ_S} we simply apply the Parseval identity (2.2.9b) and obtain

$$\int_{-\infty}^{\infty} \phi_S(t-k)\overline{\phi_S(t-\ell)}\,dt = \frac{1}{2\pi}\int_{-\infty}^{\infty} |\hat{\phi}_S(\omega)|^2 e^{-j(k-\ell)\omega}\,d\omega$$

$$= \frac{1}{2\pi}\int_{-\pi}^{\pi} e^{-j(k-\ell)\omega}\,d\omega = \delta_{k,\ell}.$$

The representation of signals in terms of the Shannon sampling (or scaling) function has another distinct feature. Suppose that $\phi_S\left(\frac{t}{a}\right)$ is used in the formula (3.1.26); then the L^2-projection mapping $L_a^S := L_a$ (for $\phi(t) = \phi_S(t)$) from L^2 into $V_{\phi_S;a,a}$ is given by

$$(L_a^S f)(t) = \sum_k \left\{\frac{1}{2\pi}\int_{-\infty}^{\infty} \hat{f}(\omega)\overline{\hat{\phi}_S(a\omega)}e^{jka\omega}\right\}\phi_S\left(\frac{t}{a}-k\right) \tag{3.2.5}$$

$$= \sum_k \left\{\frac{1}{2\pi}\int_{-\pi/a}^{\pi/a} \hat{f}(\omega)e^{jka\omega}\,d\omega\right\}\phi_S\left(\frac{t}{a}-k\right).$$

Observe that if $f(t)$ is bandlimited with bandwidth $\leq \frac{2\pi}{a}$, then

$$\frac{1}{2\pi}\int_{-\pi/a}^{\pi/a} \hat{f}(\omega)e^{jka\omega}\,d\omega = \frac{1}{2\pi}\int_{-\infty}^{\infty} \hat{f}(\omega)e^{jka\omega}\,d\omega = f(ka),$$

and so (3.2.5) implies that

$$(L_a^S f)(t) \equiv (V_a^S f)(t), \tag{3.2.6}$$

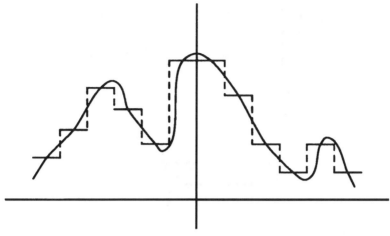

FIG. 3.2. *Piecewise constant interpolation.*

where V_a^S was introduced in (3.1.17) (see also (3.1.28) for a general $\phi(t)$). Hence, when the Shannon sampling function is used, the L^2-projection is a *generalization* of the sampling theorem from bandlimited signals to all finite-energy signals.

Observe that as a window function, $\phi_S(t) = \ell_\pi(t)$ gives "ideal" frequency-domain partition but very poor time localization. Next, let us consider the time-window function

$$M_1(t) := \chi_{[-1/2,1/2)}(t), \tag{3.2.7}$$

which, on the other hand, gives "ideal" time-domain partitioning but very poor frequency localization. For any continuous finite-energy signal $f(t)$, this window function gives rise to a piecewise constant interpolation

$$(V_a^p f)(t) := \sum_k f(ka) M_1 \left(\frac{t}{a} - k \right) \tag{3.2.8}$$

in the sense that $(V_a^p f)(t)$ is a step function with (possible) jumps at $(2k+1)\frac{a}{2}$, $k = 0, \pm 1, \ldots$ such that

$$(V_a^p f)(ka) = f(ka), \qquad k = 0, \pm 1, \ldots , \tag{3.2.9}$$

as shown in Figure 3.2.

In contrast to the Shannon (sampling) scaling function $\phi_S(t)$, however, it is clear that the representation (3.2.8) does not give perfect recovery of bandlimited signals as the representation in (3.1.17) does (by applying the sampling theorem). It only reproduces the digital samples $\{f(ka)\}$ as in (3.2.9). The merit of the mapping V_a^p in (3.2.8) over V_a^S in (3.1.17) is that it is *local* and is easier to compute and implement. Here, the concept of *localness* means that any change in the digital sample $f(k_0 a)$ does not change the signal representation function, such as $(V_a^p f)(t)$, outside a small time interval that depends on the sample point $k_0 a$, and in this case the time interval is $\left[k_0 a - \frac{a}{2}, k_0 a + \frac{a}{2} \right]$. Now, let us investigate if $M_1(t)$ is also an orthonormal scaling function. It

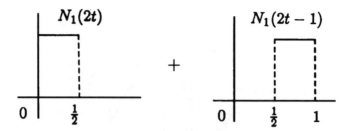

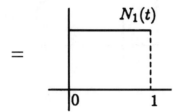

FIG. 3.3.   *Two-scale relation of $N_1(t)$.*

is obvious that $M_1(t)$ is orthonormal, since the supports of $M_1(t - k)$ and $M_1(t - \ell)$ do not even overlap for $k \neq \ell$. Unfortunately, $M_1(t)$ does *not* satisfy a two-scale relation as described by (3.2.1). To achieve such a relation, we only need to shift $M_1(t)$ half a unit to the right. This introduces the *first-order cardinal B-spline* $N_1(t)$, defined by

$$N_1(t) := M_1\left(t - \frac{1}{2}\right) = \chi_{[0,1)}(t). \qquad (3.2.10)$$

Recall from Chapter 1, section 1.2, that $N_1(t)$ is the Haar scaling function $\phi_H(t)$.

Now it is clear that $N_1(t)$ satisfies the very simple two-scale relation

$$N_1(t) = N_1(2t) + N_1(2t - 1). \qquad (3.2.11)$$

(See Figure 3.3 and recall Figure 1.9 and (1.1.14)–(1.1.15).) Hence, $N_1(t)$ *does* have the same properties of $\phi_S(t)$ as stated in Example 3.2. In other words, we have the following.

Example 3.3. The first-order cardinal B-spline $N_1(t)$ is a scaling function that is orthonormal in the sense of (3.1.24) for $a = 1$.

We will discuss the importance of the two-scale relation with *finite* (or at least fast decaying) two-scale sequences such as (3.2.11) in the next section.

3.3. Multiresolution analysis.

In signal analysis, when a lowpass filter function $\phi(t)$ is used as a sampling function (and we will say that $\phi(t)$ is a *lowpass sampling function*), say on the sample set

$$a\mathbb{Z} := \{ka:\ k = 0, \pm 1, \ldots\}, \qquad (3.3.1)$$

where $a > 0$ is the sampling period, we may consider

$$\phi\left(\frac{t}{a} - k\right) \tag{3.3.2}$$

instead of

$$\phi(t - ka) \tag{3.3.3}$$

in the signal representation (3.1.7). Indeed, in view of (3.1.4), the recovery formula (3.1.3) of the sampling theorem (see Theorem 3.1) can be formulated as

$$f(t) = \sum_{k=-\infty}^{\infty} f(ka)\phi_S\left(\frac{t}{a} - k\right). \tag{3.3.4}$$

An obvious advantage of the formulation (3.3.4) over (3.1.4) is that there is no need to use the notation $\phi_{S;a}$ in (3.1.4), but the real impact of this new signal representation is that a change in the sampling period a in (3.3.1) is directly proportional to the corresponding change in the RMS duration and inversely proportional to that in the RMS bandwidth of

$$\phi\left(\frac{t}{a} - k\right).$$

Indeed, these RMS values are given by

$$a\Delta_\phi \quad \text{and} \quad \frac{1}{a}\Delta_{\widehat{\phi}},$$

respectively. In particular, as expected, a smaller sampling period gives rise to a larger RMS bandwidth of the signal space

$$\operatorname{clos}_{L^2}\left\langle\phi\left(\frac{t}{a} - k\right) : k \in \mathbb{Z}\right\rangle.$$

Furthermore, if a lowpass sampling function $\phi(t)$ satisfies the two-scale relation (3.2.1), then by considering the scales a as integer powers of 2 it is possible to write a signal representation

$$f_n(t) = \sum_k c_{n,k}\phi(2^n t - k) \tag{3.3.5}$$

(with $a = 2^{-n}$, where n is an integer) for $n = n_1$ as another such representation for $n = n_2$ where $n_1 < n_2$. That is, a signal on a coarser grid $2^{-n_1}\mathbb{Z}$ (or lower resolution) can be represented as a signal on a finer grid $2^{-n_2}\mathbb{Z}$ (or higher resolution).

Hence, for $a = 2^n$ in (3.1.8), the signal spaces $V_{\phi,a}$ need a notational change, after a change of variables from (3.3.3) to (3.3.2), as follows:

$$V_n = V_{\phi,n} := \operatorname{clos}_{L^2}\langle\phi(2^n t - k) : k \in \mathbb{Z}\rangle. \tag{3.3.6}$$

As a consequence, if the lowpass sampling function $\phi(t)$ is a scaling function, then we have a *nested sequence*

$$\cdots \subset V_{-1} \subset V_0 \subset V_1 \subset \cdots \tag{3.3.7}$$

of (closed) subspaces V_n of the finite-energy space L^2. In fact, the nested relation in (3.3.7) can be described as follows.

Theorem 3.2. *Let $\phi(t)$ be a scaling function with two-scale sequence $\{p_k\}$ as in Definition 3.3, and let $f_n(t) \in V_n$ be given by (3.3.5), where the notation in (3.3.6) is used. Then $f_n(t) \in V_{n+1}$ and the coefficient sequences*

$$\mathbf{c}_n = \{c_{n,k}\} \quad and \quad \mathbf{c}_{n+1} = \{c_{n+1,k}\} \tag{3.3.8}$$

of the same signal

$$f_n(t) = \sum_k c_{n,k}\phi(2^n t - k) = \sum_k c_{n+1,k}\phi(2^{n+1}t - k) \tag{3.3.9}$$

in V_n and V_{n+1}, respectively, satisfy

$$c_{n+1,k} = \sum_\ell p_{k-2\ell}c_{n,\ell}. \tag{3.3.10}$$

Observe that the computation of $\mathbf{c}_{n+1}$ from $\mathbf{c}_n$ is very efficient, especially when the two-scale sequence $\{p_k\}$ is finite. It requires only two simple operations: an *upsampling* followed by a *moving average* ((MA) or discrete convolution) as shown in Figure 3.4a. Upsampling is achieved by treating the index ℓ in $c_{n,\ell}$ as an even index and inserting a zero between $c_{n,\ell}$ and $c_{n,\ell+1}$ for each ℓ as follows:

$$\begin{cases} \tilde{c}_{n,2\ell} := c_{n,\ell}, \\ \\ \tilde{c}_{n,2\ell+1} := 0. \end{cases} \tag{3.3.11}$$

(See Figure 3.4b, and recall section 1.4 of Chapter 1.)

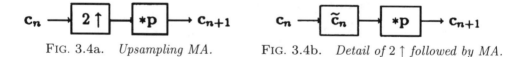

$$\mathbf{c}_n \longrightarrow \boxed{2\uparrow} \longrightarrow \boxed{*\mathbf{p}} \longrightarrow \mathbf{c}_{n+1} \qquad \mathbf{c}_n \longrightarrow \boxed{\tilde{\mathbf{c}}_n} \longrightarrow \boxed{*\mathbf{p}} \longrightarrow \mathbf{c}_{n+1}$$

FIG. 3.4a.   *Upsampling MA.*   FIG. 3.4b.   *Detail of $2\uparrow$ followed by MA.*

For example, if we use $\phi(t) = N_1(t)$, then according to (3.2.11) we have $p_0 = p_1 = 1$ and $p_k = 0$ for other indices k. Hence, (3.3.10) becomes

$$c_{n+1,2k} = c_{n+1,2k+1} = c_{n,k}, \qquad k = 0, \pm 1, \ldots . \tag{3.3.12}$$

If the two-scale sequence $\{p_k\}$ is infinite but has fast decay, then it can be truncated. Also, if the infinite sequence $\{p_k\}$ has a rational symbol, then it is the sum of two autoregressive-moving average (ARMA) filters: one operates from left to right and the other from right to left. See Example 6.3 in section 6 of Chapter 6 for a discussion of this operation.

The two-scale relation of a scaling function is the main ingredient in the following definition of MRA.

Definition 3.4. *A scaling function $\phi(t)$ is said to generate an MRA of the finite-energy space L^2, if*

(i) *the set*
$$\{\phi(t-k): \; k = 0, \pm 1, \ldots\}$$

of integer translates of $\phi(t)$ is a Riesz basis of $V_0 = V_{\phi,0} := V_\phi$ as in Definition 3.1,

and the closed subspaces $V_n = V_{\phi,n}$ of L^2, as defined in (3.3.6), satisfy

(ii) $\text{clos}_{L^2}(\cup_{n=-\infty}^\infty V_n) = L^2$.

Let us first recall that from the definition of a scaling function (in Definition 3.3), the spaces V_n also satisfy the nested property

(iii) $\cdots \subset V_{-1} \subset V_0 \subset V_1 \subset \cdots$,

as stated in (3.3.7), as well as the properties

(iv) $f(t) \in V_n \Leftrightarrow f(2t) \in V_{n+1}$ and

(v) $f(t) \in V_n \Leftrightarrow f\left(t + \frac{1}{2^n}\right) \in V_n$, for all integers n.

In addition, it follows from the definition of V_n and (iii) that

(vi) $\cap_{n=-\infty}^\infty V_n = \{0\}$.

In the early wavelet literature, statements (i)–(vi) were used together to introduce the notion of MRA. The density condition (ii) requires arbitrarily close approximation of any finite-energy function in the energy norm by functions from V_n for sufficiently large values of n, and property (vi) guarantees that if we use a very coarse grid (that is, if the signal is vastly undersampled), then the signal representation has to be a blur. Perhaps a useful descriptive notation to summarize (ii), (iii), and (vi) is the following:

$$\{0\} \leftarrow \cdots \subset V_{-1} \subset V_0 \subset V_1 \subset \cdots \rightarrow L^2. \tag{3.3.13}$$

It should be remarked that properties (iv) and (v) of an MRA have been used as basic principles in the computer vision literature.

In the previous section, we gave two examples of scaling functions, namely, $\phi_S(t)$ and $N_1(t)$. Since they are orthonormal, they already generate Riesz bases (with both upper and lower Riesz bounds $= 1$). It is also well known in the classical mathematics literature that by considering sufficiently dense sample sets, the spaces V_n generated by $\phi_S(t)$ or by $N_1(t)$ are dense in L^2, so that (ii) is satisfied. That is, each of $\phi_S(t)$ and $N_1(t)$ generates an MRA of L^2.

However, neither $\phi_S(t)$ nor $N_1(t)$ is a time-frequency window, since

$$\Delta_{\phi_S}\Delta_{\widehat{\phi_S}} = \Delta_{N_1}\Delta_{\widehat{N_1}} = \infty. \tag{3.3.14}$$

On the other hand, the Gaussian functions $h_\alpha(t)$, as given in (3.1.15), which give optimal time-frequency localization with

$$\Delta_{h_\alpha}\Delta_{\widehat{h_\alpha}} = \frac{1}{2},$$

are *not* scaling functions. To see this, recall from (3.2.1) and (3.2.2) that if $h_\alpha(t)$, $\alpha > 0$, were a scaling function, then

$$\widehat{h}_\alpha(\omega) / \widehat{h}_\alpha\left(\frac{w}{2}\right) =: P_\alpha(e^{-j\omega/2}), \qquad |\omega| < 2\pi$$

must have a 4π-periodic extension to all ω. However, a simple calculation reveals that

$$\widehat{h}_\alpha(\omega) / \widehat{h}_\alpha\left(\frac{\omega}{2}\right) = e^{-3\alpha^2\omega^2/4} \to 0. \tag{3.3.15}$$

3.4. Cardinal splines.

Since the Gaussian functions $h_\alpha(t)$, which provide optimal time-frequency localization, are not scaling functions, it is important to find scaling functions that behave like the Gaussian functions. In this section, we generalize the first-order cardinal B-spline $N_1(t)$ to the mth-order ones, denoted by $N_m(t)$, for any integer $m \geq 2$. It will be seen in section 5.4 of Chapter 5 that the B-splines $N_m(t)$ are "asymptotically optimal" time-frequency lowpass window functions in the sense that

$$\lim_{m\to\infty} \Delta_{N_m} \Delta_{\widehat{N}_m} = \frac{1}{2}. \tag{3.4.1}$$

Recall from (2.2.7) that the *uncertainty lower bound* $\frac{1}{2}$ is attained only by the Gaussian functions.

To introduce $N_m(t)$, we recall the two-scale relation (3.2.11) of $N_1(t)$ and apply the Fourier transform version (3.2.2) of this two-scale relation; namely,

$$\widehat{N}_1(\omega) = \left(\frac{1+z}{2}\right) \widehat{N}_1\left(\frac{\omega}{2}\right), \qquad z := e^{-j\omega/2}. \tag{3.4.2}$$

Taking the mth power of both sides, we then have

$$\widehat{N}_1^m(\omega) = \left(\frac{1+z}{2}\right)^m \widehat{N}_1^m\left(\frac{\omega}{2}\right), \qquad z = e^{-j\omega/2}. \tag{3.4.3}$$

Now, because the product of the Fourier transform of two functions is the Fourier transform of their convolution, we define the mth-order cardinal B-spline $N_m(t)$, $m \geq 2$, recursively, by taking convolutions of $N_1(t)$, namely,

$$N_m(t) := (N_{m-1} * N_1)(t) = \int_{-\infty}^{\infty} N_{m-1}(t-x)N_1(x)dx \tag{3.4.4}$$

$$= \int_0^1 N_{m-1}(t-x)dx,$$

and obtain

$$\widehat{N}_m(\omega) = \widehat{N}_{m-1}(\omega)\widehat{N}_1(\omega) = \widehat{N}_{m-2}(\omega)\widehat{N}_1^2(\omega) \tag{3.4.5}$$
$$= \cdots = \widehat{N}_1^m(\omega).$$

That is, the identity (3.4.3) becomes

$$\widehat{N}_m(\omega) = \left(\frac{1+z}{2}\right)^m \widehat{N}_m\left(\frac{\omega}{2}\right), \qquad z = e^{-j\omega/2} \tag{3.4.6}$$

or, equivalently, in the time domain, $N_m(t)$ satisfies the identity

$$N_m(t) = \sum_{k=0}^{m} 2^{-m+1} \binom{m}{k} N_m(2t - k). \tag{3.4.7}$$

This is the two-scale relation of the scaling function $N_m(t)$. Now, since

$$\widehat{N}_1(\omega) = \int_0^1 e^{-j\omega t} dt = \frac{1 - e^{-j\omega}}{j\omega},$$

it follows from (3.4.5) that

$$\widehat{N}_m(\omega) = \left(\frac{1 - e^{-j\omega}}{j\omega}\right)^m \tag{3.4.8}$$

and

$$|\widehat{N}_m(\omega)| = \left|\frac{\sin(\frac{\omega}{2})}{\frac{\omega}{2}}\right|^m. \tag{3.4.9}$$

Hence, for each $m \geq 1$, $N_m(t)$ is a first-order Butterworth filter:

$$|\widehat{N}_m(0)| = 1, \quad \left[\frac{d}{d\omega}|\widehat{N}_m(\omega)|\right]_{\omega=0} = 0, \quad \left[\frac{d^2}{d\omega^2}|\widehat{N}_m^2(\omega)|\right]_{\omega=0} \neq 0. \tag{3.4.10}$$

See (2.3.3)–(2.3.4). The graphs of the lowpass windows $N_m(t)$ along with their filter characteristics $|\widehat{N}_m(\omega)|$ are shown in Figures 3.5–3.8 for $m = 1, \ldots, 4$, respectively.

Observe that the *side lobes* of $|\widehat{N}_m(\omega)|$ diminish with increasing values of m. By applying (3.4.9), the *side-lobe/main-lobe power ratio* of the filter characteristic $|\widehat{N}_m(\omega)|$ can be formulated as

$$(S/M)(\widehat{N}_m) := \frac{\int_{2\pi}^{\infty} |\widehat{N}_m(\omega)|^2 \, d\omega}{\int_0^{2\pi} |\widehat{N}_m(\omega)|^2 \, d\omega}$$
$$= \frac{\int_\pi^{\infty} \left(\frac{\sin\omega}{\omega}\right)^{2m} d\omega}{\int_0^\pi \left(\frac{\sin\omega}{\omega}\right)^{2m} d\omega}. \tag{3.4.11}$$

These values (up to $m = 6$) are given in Table 3.1.

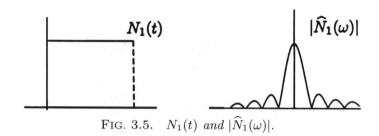

FIG. 3.5. $N_1(t)$ and $|\widehat{N}_1(\omega)|$.

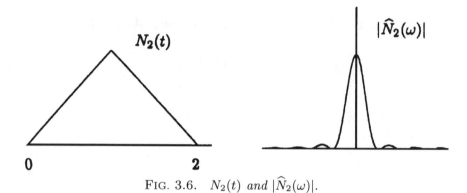

FIG. 3.6. $N_2(t)$ and $|\widehat{N}_2(\omega)|$.

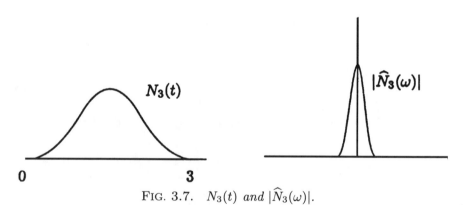

FIG. 3.7. $N_3(t)$ and $|\widehat{N}_3(\omega)|$.

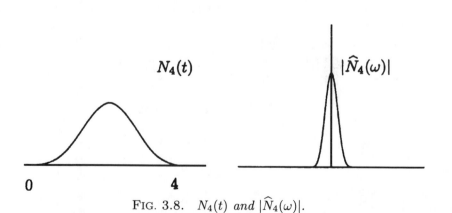

FIG. 3.8. $N_4(t)$ and $|\widehat{N}_4(\omega)|$.

TABLE 3.1.   *Side-lobe/main-lobe power ratios $(S/M)(\widehat{N}_m)$ in $-dB$.*

m	1	2	3	4	5	6
$(S/M)(\widehat{N}_m)$ in $- dB$	9.681	25.299	39.021	52.408	65.704	78.974

In many engineering applications such as in antenna design, however, the emphasis is not on the power contents but on the maximum values of the main lobe and its two neighboring side lobes. For the function $(\sin \omega/\omega)^m$ in (3.4.9), the main lobe is located in the interval $|\omega| \leq \pi$, where the maximum of the function is 1, while the neighboring side lobes are in $\pi \leq |\omega| \leq 2\pi$. It is clear that the maximum there is attained at the (unique) solution $\widehat{\omega}$ of the equation

$$\tan \omega = \omega, \qquad \pi \leq |\omega| \leq 2\pi.$$

Hence, the *maxima side-lobe/main-lobe ratio* of the filter characteristic of $N_m(t)$ is given by

$$(MS/MM)(\widehat{N}_m) := \frac{\max\{|\widehat{N}_m(\omega)|\colon \pi \leq |\omega| \leq 2\pi\}}{\max\{|\widehat{N}_m(\omega)|\colon |\omega| \leq \pi\}}. \qquad (3.4.12)$$

These values (up to $m = 6$) are given in Table 3.2.

TABLE 3.2.   *Maxima side-lobe/main-lobe ratios in $-dB$.*

m	1	2	3	4	5	6
$(MS/MM)(\widehat{N}_m)$ in $-dB$	6.6307	13.261	19.892	25.523	33.154	39.784

From its definition (3.2.10), we see that $N_1(t)$ is a piecewise constant function. Let

$$\pi_{m-1} := \langle 1, t, \ldots, t^{m-1} \rangle \qquad (3.4.13)$$

denote the collection of all polynomials of degree $\leq m - 1$ (or order m), and

$$C^k = C^k(-\infty, \infty) := \{f(t)\colon f(t), \ldots, f^{(k)}(t) \text{ continuous}\}$$

denote the class of all k-times continuously differentiable functions on $\mathbb{R} = (-\infty, \infty)$. Since the indefinite integral of a polynomial increases the degree of the polynomial by one and the integral convolution operation is a smoothing process, it follows from formulation (3.4.4) that $N_m(t)$ is a piecewise polynomial function of degree $m - 1$ in C^{m-2}. To list the important properties of $N_m(t)$, we need the following notations:

$$\text{supp } f := \{t\colon f(t) \neq 0\}, \qquad (3.4.14)$$

$$f|_{[a,b]} := \text{ the restriction of } f(t) \text{ on } [a, b], \qquad (3.4.15)$$

called support of $f(t)$ and restriction of $f(t)$, respectively.

Theorem 3.3. *For each $m \geq 2$, the mth-order cardinal B-spline $N_m(t)$, as defined in (3.4.4), has the following properties:*

(i) $N_m(t) \in C^{m-2}$;

(ii) $N_m|_{[k,k+1]} \in \pi_{m-1}, k = 0, \pm 1, \ldots$;

(iii) supp $N_m = [0, m]$;

(iv) $N_m(t) > 0$ *for* $0 < t < m$;

(v) $\sum_{k=-\infty}^{\infty} N_m(t - k) = 1$ *all* t;

(vi) $\int_{-\infty}^{\infty} N_m(x)dx = 1$;

(vii) $N_m'(t) = N_{m-1}(t) - N_{m-1}(t - 1)$;

(viii) $N_m(t)$ *can be computed from* $N_{m-1}(t)$ *by using the identity*

$$N_m(t) = \frac{t}{m-1} N_{m-1}(t) + \frac{m-t}{m-1} N_{m-1}(t - 1);$$

(ix) $N_m(t)$ *is symmetric relative to its center*

$$t_m^* := \frac{m}{2}; \tag{3.4.16}$$

(x) $N_m(t)$ *is a scaling function, with two-scale relation given by (3.4.7) that generates an MRA* $\{V_n^m\}$ *of* L^2, *where*

$$V_n^m := \text{clos}_{L^2}\langle N_m(2^n t - k): k \in \mathbb{Z}\rangle. \tag{3.4.17}$$

In general, any function $f(t)$ satisfying

$$f(t) \in C^{m-2} \quad \text{and} \quad f|_{[ka,(k+1)a]} \in \pi_{m-1}, \qquad k = 0, \pm 1, \ldots \tag{3.4.18}$$

is called an mth-order cardinal spline with knot sequence $a\mathbb{Z}$, where $a > 0$. The reason for $N_m(t)$ to be called a B-spline (or **basic** spline) is that every $f(t)$ satisfying (3.4.18) can be written as

$$f(t) = \sum_{k=-\infty}^{\infty} c_k N_m\left(\frac{t}{a} - k\right) \tag{3.4.19}$$

for some constants c_k. In particular, since $N_m(t)$ has finite time duration (or compact support), we have

$$f(t) \in L^2 \quad \text{if and only if} \quad \{c_k\} \in \ell^2. \tag{3.4.20}$$

To determine the coefficient sequence $\{c_k\}$ from $f(t)$ in (3.4.19), we need the following notion of dual B-splines.

Definition 3.5. *For each integer $m \geq 1$, the mth-order cardinal dual B-spline $\widetilde{N}_m(t)$ is the mth-order cardinal spline, with the same knot sequence $\mathbb{Z}$ as $N_m(t)$ that satisfies the dual condition*

$$\langle N_m(t-k), \widetilde{N}_m(t-\ell) \rangle := \int_{-\infty}^{\infty} N_m(t-k)\widetilde{N}_m(t-\ell)dt = \delta_{k,\ell}, \qquad k, \ell \in \mathbb{Z}.$$
(3.4.21)

Hence, by applying (3.4.21), the coefficients c_k in (3.4.19) are given by

$$c_k = \frac{1}{a} \int_{-\infty}^{\infty} f(t)\widetilde{N}_m\left(\frac{t}{a} - k\right) dt.$$
(3.4.22)

Since $N_m(t)$ is symmetric with respect to its center, it follows from (3.4.16) and (3.4.21) that the center $\tilde{t}_m^*$ of $\widetilde{N}_m(t)$ is also

$$\tilde{t}_m^* = \frac{m}{2}.$$
(3.4.23)

Hence, for each index k, the coefficient c_k of $f(t)$ in (3.4.19) gives the local information of $f(t)$ around the time location

$$t = \left(k + \frac{m}{2}\right)a.$$

This information is revealed in the time-frequency window

$$\left[\left(k + \frac{m}{2}\right)a - \Delta_{\widetilde{N}_m}, \quad \left(k + \frac{m}{2}\right)a + \Delta_{\widetilde{N}_m}\right] \times \left[-\Delta_{\widehat{\widetilde{N}}_M}, \Delta_{\widehat{\widetilde{N}}_m}\right].$$
(3.4.24)

The graphs of $\widetilde{N}_m(t)$ along with their (lowpass) filter characteristics $|\widehat{\widetilde{N}}_m(\omega)|$ are shown in Figures 3.9–3.11 for $m = 2, 3, 4$, respectively. Observe that $\widetilde{N}_1(t) = N_1(t)$.

More details on duals will be discussed in Chapters 5 and 6. (For instance, see Chapter 5 for a theoretical discussion and Chapter 6 for applications to computation and implementations.)

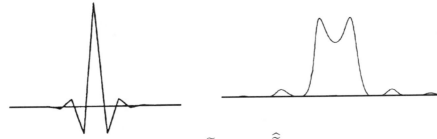

FIG. 3.9. $\widetilde{N}_2(t)$ and $|\widehat{\widetilde{N}}_2(\omega)|$.

FIG. 3.10. $\widetilde{N}_3(t)$ and $|\widehat{\widetilde{N}}_3(\omega)|$.

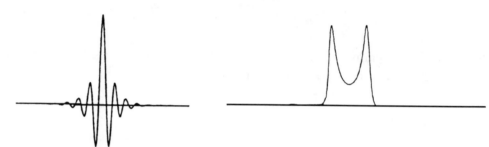

FIG. 3.11. $\widetilde{N}_4(t)$ and $|\widehat{\widetilde{N}}_4(\omega)|$.

CHAPTER 4

Orthonormal Wavelets

For signal analysis, a wavelet $\psi(t)$ is considered to be a bandpass window function that stops at least the zero frequency; namely,

$$\int_{-\infty}^{\infty} \psi(t)dt = \widehat{\psi}(0) = 0.$$

This property enables the integral wavelet transform (IWT), or continuous wavelet transform (CWT), with $\psi(t)$ as the analyzing wavelet, to annihilate the flat (i.e., near-constant) segments of the analog signal and thus provide a better understanding of its details. If $\psi(t)$ has vanishing moments of a higher order, meaning that

$$\int_{-\infty}^{\infty} t^k \psi(t)dt = 0, \qquad k = 0, \ldots, m-1,$$

for some integer $m \geq 2$, then even the "smooth polynomial" segments are annihilated, and the IWT can better reveal the details of the signal on each octave band by considering different scales a (see section 6.5 in Chapter 6).

In the construction of wavelets, we will take full advantage of the multi-scale structure of the notion of MRA as introduced in Chapter 3. Now, suppose that such a multiresolution analysis (MRA) $\{V_n\}$ of L^2 is generated by some scaling function $\phi(t)$. Then if the integer translates of $\phi(t)$ locally reproduce all polynomials of degree $\leq m-1$ and if $\psi(t)$ is constructed to be orthogonal to all the integer translates of $\phi(t)$, then in each nth octave band, the wavelets $\psi(2^n t - k)$ are orthogonal to V_n, and hence to all polynomials in π_{m-1}, at least locally.

We remark in passing that if $\phi(t)$ has compact support (i.e., finite time duration) in the sense that supp ϕ, as defined in (3.4.14), is a bounded set, then the property of local reproduction of π_{m-1} by integer translates $\phi(t-k)$ of $\phi(t)$ is equivalent to the mth order of approximation, say $O(a^m)$, from the scale space

$$\left\langle \phi\left(\frac{t}{a} - k\right) : k \in \mathbb{Z} \right\rangle.$$

In section 4.1, we will clarify this orthogonality structure and introduce some of the basic tools concerning orthogonality. In the remaining sections of this chapter, we will study the well-known orthonormal wavelets based on the MRA and orthogonality structure: first the wavelets of Haar, Shannon, and Meyer in section 4.2, then the spline wavelets of Battle-Lemarié and Strömberg in section 4.3, and finally the Daubechies wavelets in section 4.4.

4.1. Orthogonal wavelet spaces.

Let $\phi(t)$ be a (lowpass) scaling function that generates an MRA

$$\{0\} \leftarrow \cdots \subset V_{-1} \subset V_0 \subset V_1 \subset \cdots \rightarrow L^2 \tag{4.1.1}$$

of L^2. For each integer n, since V_{n-1} is a proper subspace of V_n, we have a nontrivial orthogonal complementary subspace W_{n-1} of V_n relative to V_{n-1}. That is, $W_{n-1} \subset V_n$ and

$$V_n = V_{n-1} + W_{n-1}, \quad W_{n-1} \perp V_{n-1}. \tag{4.1.2}$$

We will use the notation

$$V_n = V_{n-1} \oplus W_{n-1} \tag{4.1.3}$$

to describe (4.1.2). It is clear from (4.1.1) and (4.1.2) that

$$W_n \perp W_k \quad \text{all } n \neq k, \tag{4.1.4}$$

and by repeated applications of (4.1.3) we have, for all $n \in \mathbb{Z}$ and $M < n$,

$$\begin{aligned}
V_n &= V_{n-1} \oplus W_{n-1} = (V_{n-2} \oplus W_{n-2}) \oplus W_{n-1} \\
&= \cdots = V_{n-M} \oplus (W_{n-M} \oplus \cdots \oplus W_{n-1}).
\end{aligned} \tag{4.1.5}$$

Hence, in view of (4.1.1), the finite-energy signal space L^2 can be decomposed as an orthogonal sum of the subspaces W_n; namely,

$$L^2 = \bigoplus_{n=-\infty}^{\infty} W_n. \tag{4.1.6}$$

In the following, we will see that all of these subspaces W_n are generated by a single function $\psi(t)$ that can be used as an analyzing wavelet to define the IWT. For this reason, (4.1.6) will be called a *wavelet decomposition* of L^2. In applications to signal analysis, however, the finite decomposition in (4.1.5) is preferred over the infinite decomposition in (4.1.6). The component of a decomposed signal in W_n will be called its *nth octave*, and the one in V_{n-M} its *blur* (or DC) component.

If a function $\psi(t) \in L^2$, with sufficiently fast decay at infinity, has been chosen to generate the zeroth-octave subspace W_0, in the sense that

$$W_0 = \text{clos}_{L^2} \langle \psi(t - k) : k \in \mathbb{Z} \rangle,$$

then it is clear that $\psi(t)$ also generates all the other octave spaces W_n by scaling in powers of 2; namely,

$$W_n = \mathrm{clos}_{L^2}\langle \psi(2^n t - k) \colon k \in \mathbb{Z}\rangle, \qquad n \in \mathbb{Z}. \qquad (4.1.7)$$

If, in addition, the scaling function $\phi(t)$ locally reproduces all polynomials in π_{m-1} in the sense that some real constants $a_{k,n}$ exist such that

$$t^k = \sum_{n=-\infty}^{\infty} a_{k,n}\phi(t - n), \qquad k = 0, \ldots, m - 1 \qquad (4.1.8)$$

hold on any bounded interval of t, then since $W_0 \perp V_0$ we have

$$\int_{-\infty}^{\infty} t^k \psi(t)\,dt = 0, \qquad k = 0, \ldots, m - 1, \qquad (4.1.9)$$

provided, of course, that the decay of $\psi(t)$ at infinity is fast enough that (4.1.9) makes sense. Consequently, since $W_n \perp V_n$ for all $n \in \mathbb{Z}$, each octave component of a signal reveals the detail of the signal in the corresponding frequency band.

Recall that a scaling function $\phi(t)$ that generates an MRA $\{V_n\}$ of L^2 must be stable. This means that it satisfies the Riesz condition:

$$A\|\{c_k\}\|_{\ell_2}^2 \le \left\|\sum_k c_k \phi(t - k)\right\|^2 \le B\|\{c_k\}\|_{\ell_2}^2 \quad \text{all} \quad \{c_k\} \in \ell^2, \qquad (4.1.10)$$

where $0 < A \le B < \infty$. (See (3.1.14) in Chapter 3.) This set of inequalities can be shown to be equivalent to the following condition on the Fourier transform $\widehat{\phi}(\omega)$ of $\phi(t)$, with the same constants A and B; namely,

$$A \le \sum_{k=-\infty}^{\infty} |\widehat{\phi}(\omega + 2\pi k)|^2 \le B, \quad \text{a.e.} \qquad (4.1.11)$$

We will use (4.1.10) and (4.1.11) interchangeably. In particular, we will apply (4.1.11) to orthonormalize the Riesz basis

$$\{\phi(t - k) \colon k \in \mathbb{Z}\} \qquad (4.1.12)$$

of V_0. To do so, we need the following orthogonality criterion.

Theorem 4.1. *A function $\eta(t) \in L^2$ is orthonormal in the sense that*

$$\langle \eta(t - k), \eta(t - \ell)\rangle := \int_{-\infty}^{\infty} \eta(t - k)\overline{\eta(t - \ell)}\,dt = \delta_{k,\ell}, \qquad k, \ell \in \mathbb{Z} \quad (4.1.13)$$

if and only if its Fourier transform $\widehat{\eta}(\omega)$ satisfies

$$\sum_{n=-\infty}^{\infty} |\widehat{\eta}(\omega + 2\pi n)|^2 = 1, \quad \text{a.e.} \qquad (4.1.14)$$

The proof of this theorem is a direct application of the Parseval identity (2.2.9b). Indeed, it follows from

$$
\int_{-\infty}^{\infty} \eta(t-k)\overline{\eta(t-\ell)}\, dt = \frac{1}{2\pi}\int_{-\infty}^{\infty} |\hat{\eta}(\omega)|^2 e^{-j(k-\ell)\omega}\, d\omega
$$

$$
= \frac{1}{2\pi}\sum_{n=-\infty}^{\infty}\int_{2\pi n}^{2\pi(n+1)} |\hat{\eta}(\omega)|^2 e^{-j(k-\ell)\omega}\, d\omega
$$

$$
= \frac{1}{2\pi}\int_{0}^{2\pi}\left\{\sum_{n=-\infty}^{\infty} |\hat{\eta}(\omega+2\pi n)|^2\right\} e^{-j(k-\ell)\omega}\, d\omega
$$

that (4.1.13) and (4.1.14) are equivalent.

Now, if $\phi(t)$ generates a Riesz basis so that (4.1.11) is satisfied, then the function $\phi^{\perp}(t)$ whose Fourier transform is given by

$$
\hat{\phi}^{\perp}(\omega) = \frac{\hat{\phi}(\omega)}{\left(\sum_{k=-\infty}^{\infty} |\hat{\phi}(\omega+2\pi k)|^2\right)^{1/2}} \tag{4.1.15}
$$

clearly satisfies (4.1.14). Hence, by Theorem 4.1, we see that the family

$$
\{\phi^{\perp}(t-k): k = 0, \pm 1, \ldots\} \tag{4.1.16}
$$

is orthonormal. Later, we will call (4.1.15) an *orthonormalization process* of the Riesz basis in (4.1.12).

Observe that if $\phi(t)$ generates an MRA $\{V_n\}$ of L^2, then $\phi^{\perp}(t)$ also generates the same MRA. Let $\{p_k\}$ be its two-scale sequence; that is, $\phi^{\perp}(t)$ satisfies the identity

$$
\phi^{\perp}(t) = \sum_{k} p_k \phi^{\perp}(2t-k). \tag{4.1.17}
$$

In the following, we will see that the finite-energy function

$$
\psi(t) := \sum_{k} (-1)^k p_{1-k} \phi^{\perp}(2t-k) \tag{4.1.18}
$$

generates the orthogonal complementary subspaces W_n relative to the MRA as described by (4.1.2) in the sense of (4.1.7). (Here, since we only consider real-valued scaling functions $\phi(t)$, $\phi^{\perp}(t)$ is also real valued, and, hence, the coefficients p_k in (4.1.17) are real constants. If complex p_k's are considered, then p_{1-k} in (4.1.18) must be replaced by its complex conjugate $\overline{p}_{1-k}$.)

To verify that $\psi(t)$, defined in (4.1.18), is in W_0, we simply observe that, for each $\ell \in \mathbb{Z}$,

$$
\langle \phi(t-\ell), \psi(t)\rangle = \sum_{k,n}(-1)^n p_k p_{1-n}\langle\phi^{\perp}(2t-2\ell-k), \phi^{\perp}(2t-n)\rangle
$$

$$
= \frac{1}{2}\sum_{k,n}(-1)^n p_k p_{1-n}\delta_{2\ell+k,n} = \frac{1}{2}\sum_{k}(-1)^{2\ell+k} p_k p_{1-2\ell-k}
$$

$$
= \frac{1}{2}\sum_{k}(-1)^{2\ell+k-1} p_{1-2\ell-k}p_k,
$$

which must be 0 (since the last two quantities are negative of each other).

To ensure that $\psi(t)$ generates all of W_0, we write down the following identity that describes the orthogonal decomposition $V_1 = V_0 \oplus W_0$; namely,

$$\phi(2t - \ell) = \sum_k \left\{ \frac{1}{2} p_{\ell-2k} \phi(t - k) + \frac{1}{2}(-1)^\ell p_{2k-\ell+1} \psi(t - k) \right\}. \qquad (4.1.19)$$

This identity is called the *decomposition relation* of $\phi(t)$ and $\psi(t)$.

Finally, it can also be verified that the family of all integer translates of $\psi(t)$ is orthonormal. We summarize these properties of $\psi(t)$ in the following.

Theorem 4.2. *Let $\phi(t)$ be a scaling function that generates an MRA $\{V_n\}$ of L^2, and let $\phi^\perp(t)$ be its orthonormalization as defined in (4.1.15). Also, let $\{p_k\}$ be the two-scale sequence of $\phi^\perp(t)$ and $\psi(t)$ be defined as in (4.1.18) by using $\phi^\perp(t)$ and $\{p_k\}$. Then $\psi(t)$ generates the orthogonal complementary subspaces W_n relative to the MRA $\{V_n\}$ in the sense of (4.1.7) and (4.1.3). Furthermore, the family*

$$\psi_{j,k}(t) := 2^{j/2} \psi(2^j t - k), \qquad j, k \in \mathbb{Z} \qquad (4.1.20)$$

is an orthonormal basis of the finite-energy space L^2.

The normalization by the factor $2^{j/2}$ in (4.1.20) ensures that

$$\|\psi_{j,k}\| = \|\psi\| \quad \text{for all } j, k \in \mathbb{Z}. \qquad (4.1.21)$$

Hence, the last statement in the above theorem follows from (4.1.6), (4.1.21), and the fact that the integer translates of $\psi(t)$ are orthonormal.

4.2. Wavelets of Haar, Shannon, and Meyer.

In order to emphasize the importance of Theorem 4.2, we give the following formal definition of orthonormal wavelets.

Definition 4.1. *A function $\psi(t) \in L^2$ is called an* orthonormal wavelet *if the collection of functions $\psi_{j,k}(t)$, $j, k \in \mathbb{Z}$, as defined in (4.1.20), is an orthonormal basis of L^2.*

The outline given in Theorem 4.2 can be followed to construct an orthonormal wavelet from any given scaling function that generates an MRA of L^2. Without going into any detail, we remark that for any family $\{\psi_{j,k}(t)\}$ in Definition 4.1 to be an orthonormal basis of L^2, the function $\psi(t)$ itself must necessarily have zero mean, and, hence, $\psi(t)$ is indeed a wavelet.

Example 4.1 (the Haar wavelet).

Recall from section 3.3 in Chapter 3 that the first-order B-spline $N_1(t)$ is a scaling function that generates an MRA of L^2. Since $N_1(t)$ is already orthonormal (as stated in Example 3.3), we have $N_1^\perp(t) = N_1(t)$, so that by using the two-scale coefficients

$$p_0 = p_1 = 1 \quad \text{and} \quad p_k = 0 \quad \text{for} \quad k \neq 0, 1 \qquad (4.2.1)$$

of $N_1(t)$, as given in (3.2.11), we can apply (4.1.18) to find the orthonormal Haar wavelet

$$\psi_H(t) := N_1(2t) - N_1(2t - 1) \tag{4.2.2}$$

shown in Figure 4.1. To decompose $N_1(2t - \ell) \in V_1$ as the sum of a function in V_0 and another in W_0, we apply (4.1.19) and obtain

$$N_1(2t) = \frac{1}{2}N_1(t) + \frac{1}{2}\psi_H(t), \tag{4.2.3a}$$

$$N_1(2t - 1) = \frac{1}{2}N_1(t) - \frac{1}{2}\psi_H(t). \tag{4.2.3b}$$

(See Figure 4.2 and recall (1.1.11) with $\phi_H(t) = N_1(t)$ and Figure 1.8.) Note that (4.2.3a) and (4.2.3b) can be used to decompose $N_1(2t - \ell)$ for all even ℓ and all odd ℓ, respectively. Recall the formulation of (1.1.13) by using (1.1.11)–(1.1.12).

Example 4.2 (the Shannon wavelet).

The Shannon sampling function $\phi_S(t)$ defined in (3.1.5) is also an orthonormal scaling function that generates an MRA of L^2. (See Example 3.2 and section 3.2 in Chapter 3.) Hence, we have $\phi_S^\perp(t) = \phi_S(t)$, and the two-scale sequence of $\phi_S(t)$ in (3.2.4) can be used in (4.1.18) to yield the orthonormal Shannon wavelet:

$$\psi_S(t) := \sum_k (-1)^k p_{1-k} \phi_S(2t - k) \tag{4.2.4}$$

$$= -\phi_S(2t - 1) + \sum_k (-1)^k \frac{2}{(2k + 1)\pi} \phi_S(2t - 2k - 2)$$

$$= -\phi_S(2t - 1) + \left\{ \phi_S\left(t - \frac{1}{2}\right) - \phi_S\left(2\left(t - \frac{1}{2}\right)\right) \right\}$$

$$= \phi_S\left(t - \frac{1}{2}\right) - 2\phi_S(2t - 1)$$

$$= \frac{\sin 2\pi t - \cos \pi t}{\pi(t - \frac{1}{2})}.$$

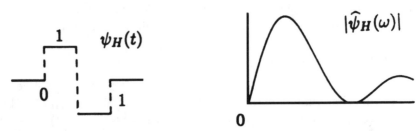

FIG. 4.1.   *The Haar wavelet $\psi_H(t)$ and its bandpass filter characteristic.*

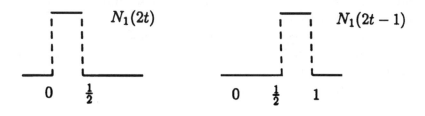

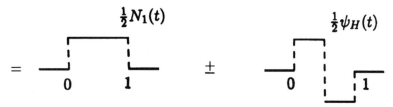

FIG. 4.2. *Decomposition relation of $N_1(t)$ and $\psi_H(t)$.*

Here, t is replaced by $t - \frac{1}{2}$ in (3.2.4) to give the second equality, and (3.1.5) is applied to yield the final expression. See Figure 4.3 for the graph of $\psi_S(t)$ and compare it with that of the bandpass filter $b_{\omega_{n-1},\omega_n}(t)$ in Figure 2.4b. It seems as if the graph of $\psi_S(t)$ is only the negative of that in Figure 2.4b. Indeed, comparing the formula of $\psi_S(t)$ in (4.2.4) with the formula of $b_{\pi,2\pi}(t)$ in (2.1.7), we have

$$\psi_S(t) = -b_{\pi,2\pi}\left(t - \frac{1}{2}\right), \tag{4.2.5}$$

and, hence,

$$\widehat{\psi}_S(\omega) = -e^{-j\omega/2}\chi_{[-2\pi,-\pi)\cup(\pi,2\pi]}(\omega). \tag{4.2.6}$$

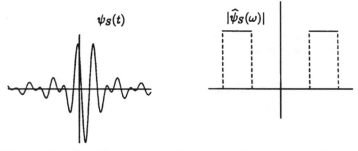

FIG. 4.3. *The Shannon wavelet and its filter characteristic.*

If we plot the filter characteristics of $\phi_S(2t)$, $\phi_S(t)$, and $\psi_S(t)$ only on the positive frequency axis, we can really appreciate the meaning of multiresolution and wavelet analysis as a mathematical tool for separating an RMS bandlimited signal into a low-frequency (DC or blur) component and a high-frequency (AC or wavelet octave) component. In the extreme case as illustrated by the Shannon scaling function $\phi_S(t)$ and wavelet $\psi_S(t)$ in Figure 4.4, the RMS bandwidth is replaced by the standard bandwidth. See (2.1.19) for a comparison when bandlimited signals are considered.

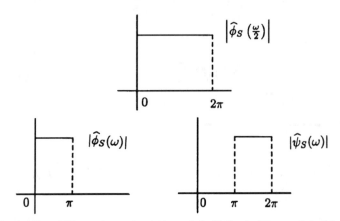

FIG. 4.4.   *Filter characteristics of $\phi_S(2t)$, $\phi_S(t)$, and $\psi_S(t)$.*

Example 4.3 (the Meyer wavelets).

The reason for the infinite RMS duration of the Shannon scaling function $\phi_S(t)$ is its slow decay at infinity. This, in turn, is a consequence of the fact that the Fourier transform $\widehat{\phi}_S(\omega)$ of $\phi_S(t)$ has jump discontinuities at $\omega = \pm\pi$. Hence, to modify $\phi_S(t)$ so as to achieve finite RMS duration, we need to *smooth* $\widehat{\phi}_S(\omega)$ at $\omega = \pm\pi$. In the following, we allow the *transition bands* of the ideal lowpass filter $\widehat{\phi}_S(\omega)$ to be

$$\left[-\frac{4}{3}\pi, -\frac{2}{3}\pi\right] \cup \left[\frac{2}{3}\pi, \frac{4}{3}\pi\right] \tag{4.2.7}$$

for this smoothing purpose. The result is a class of orthonormal Meyer scaling functions. In Figure 4.5, we compare the filter characteristic of $\phi_S(t)$ and that of some desirable Meyer scaling function $\phi_{M;N}(t)$, which we will call the Nth-order orthonormal *Meyer scaling function.*

Depending on the required order of smoothness of $\widehat{\phi}_{M;N}(\omega)$, we select a nonnegative integer N and introduce a *corner-smoothing function* $s_N(x) \in C^N$ that satisfies

$$\begin{cases} s_N(x) = 0 & \text{for} \quad x \le 0, \\ s_N(x) + s_N(1-x) = 1 & \text{for} \quad 0 \le x \le 1, \\ s_N(x) = 1 & \text{for} \quad x \ge 1. \end{cases} \tag{4.2.8}$$

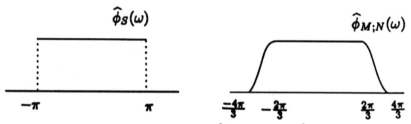

FIG. 4.5. *Comparison of* $\widehat{\phi}_{M;N}(\omega)$ *with* $\widehat{\phi}_S(\omega)$.

See Figure 4.6 for the graphs of $s_N(x)$. For instance, we may choose

$$s_0(x) = \begin{cases} 0 & \text{for} \quad x \leq 0, \\ x & \text{for} \quad 0 \leq x \leq 1, \\ 1 & \text{for} \quad x \geq 1; \end{cases} \tag{4.2.9}$$

$$s_1(x) = \begin{cases} 0 & \text{for} \quad x \leq 0, \\ 2x^2 & \text{for} \quad 0 \leq x \leq \frac{1}{2}, \\ 1 - 2(1-x)^2 & \text{for} \quad \frac{1}{2} \leq x \leq 1, \\ 1 & \text{for} \quad x \geq 1. \end{cases} \tag{4.2.10}$$

Without going into any detail, we remark that higher-order corner smoothing functions $s_N(x)$ can be easily constructed by using piecewise polynomials (or splines) of higher degrees with more knots in the interval $[0,1]$. In fact, even a C^∞ function $s_\infty(x)$ can be constructed by making use of the functions $e^{-|x-a|^{-1}}$ to satisfy (4.2.8).

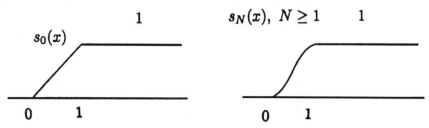

FIG. 4.6. *Corner-smoothing functions.*

Now, with the aid of any corner-smoothing function $s_N(x)$, we introduce the function

$$\cos\left(\frac{\pi}{2} s_N\left(\frac{3}{2\pi}|\omega| - 1\right)\right) \tag{4.2.11}$$

to fill in the transition bands in (4.2.7) in order to complete the graph of $\widehat{\phi}_{M;N}(\omega)$ in Figure 4.5. That is, the orthonormal Meyer scaling function

$\phi_{M;N}(t)$ of order N is defined as the inverse Fourier transform of

$$\widehat{\phi}_{M;N}(\omega) := \begin{cases} 0 & \text{for } |\omega| \geq \frac{4\pi}{3}, \\ \cos\left(\frac{\pi}{2}s_N\left(\frac{3}{2\pi}|\omega| - 1\right)\right) & \text{for } \frac{2\pi}{3} \leq |\omega| \leq \frac{4\pi}{3}, \\ 1 & \text{for } |\omega| \leq \frac{2\pi}{3}. \end{cases} \quad (4.2.12)$$

Since $s_N(x) \in C^N$, it is clear that $\widehat{\phi}_{M;N}(\omega) \in C^N$ also. The reason for taking the composition of $s_N(x)$ with the cosine function in (4.2.11)–(4.2.12) is to ensure that

$$|\widehat{\phi}_{M;N}(\omega)|^2 + |\widehat{\phi}_{M;N}(\omega + 2\pi)|^2 = 1 \quad \text{for} \quad \frac{2\pi}{3} \leq \omega \leq \frac{4\pi}{3}.$$

Since the support of $\widehat{\phi}_{M;N}(\omega + 2\pi k)$ for any $k \neq 0, 1$ is disjoint with the interval $\left(\frac{2\pi}{3}, \frac{4\pi}{3}\right)$, we have

$$\sum_{k=-\infty}^{\infty} |\widehat{\phi}_{M;N}(\omega + 2\pi k)|^2 = 1 \quad \text{all} \quad \omega. \quad (4.2.13)$$

Hence, by Theorem 4.1, we see that the family

$$\{\phi_{M;N}(t - k): \ k = 0, \pm 1, \ldots\}$$

is orthonormal.

To show that $\phi_{M;N}(t)$ is a scaling function, we simply observe that $\widehat{\phi}_{M;N}(\omega) = \widehat{\phi}_{M;N}(\frac{\omega}{2})\widehat{\phi}_{M;N}(\omega)$ in supp $\widehat{\phi}_{M;N}$ and follow Example 3.2 in section 3.2 of Chapter 3 to conclude that the two-scale symbol $P(z)$ of $\phi_{M;N}(t)$, for $z = e^{-j\omega/2}$, is the Fourier series of the 4π-periodic extension of $\widehat{\phi}_{M;N}(\omega)$. In other words, we have

$$\frac{1}{2} \sum_{k=-\infty}^{\infty} p_k e^{-jn\omega/2} = \sum_{k=-\infty}^{\infty} \widehat{\phi}_{M;N}(\omega + 4\pi k), \quad (4.2.14)$$

so that

$$p_n = \sum_{k=-\infty}^{\infty} \frac{1}{2\pi} \int_{-2\pi}^{2\pi} e^{jn\omega/2}\widehat{\phi}_{M;N}(\omega + 4\pi k)d\omega \quad (4.2.15)$$

$$= \frac{1}{2\pi} \int_{-\infty}^{\infty} e^{jn\omega/2}\widehat{\phi}_{M;N}(\omega)d\omega$$

$$= \frac{1}{\pi} \int_{0}^{2\pi} \cos\frac{n\omega}{2}\widehat{\phi}_{M;N}(\omega)d\omega$$

$$= \frac{2\sin\frac{n\pi}{3}}{3\frac{n\pi}{3}} + \frac{1}{\pi} \int_{2\pi/3}^{4\pi/3} \cos\frac{n\omega}{2}\cos\left(\frac{\pi}{2}s_N\left(\frac{3}{2\pi}\omega - 1\right)\right)d\omega.$$

Hence, by Theorem 4.2, the Nth-order Meyer wavelet $\psi_{M;N}(t)$ can be computed by putting the above values of p_n into (4.1.18), yielding

$$\psi_{M;N}(t) = \sum_{k=-\infty}^{\infty} (-1)^k p_{1-k} \phi_{M;N}(2t - k) = \sum_{k=-\infty}^{\infty} (-1)^{k+1} p_k \phi_{M;N}(2t + k - 1).$$

$$(4.2.16)$$

We also remark that, as a consequence of (4.2.15), we have

$$\phi_{M;N}\left(\frac{n}{2}\right) = p_n \quad \text{all} \quad n \in \mathbb{Z} \qquad (4.2.17)$$

That is, just as the Shannon sampling function $\phi_S(t)$, the Meyer scaling functions are "interpolatory" in the sense that

$$\phi_{M;N}(t) = \sum_{k=-\infty}^{\infty} \phi_{M;N}\left(\frac{k}{2}\right) \phi_{M;N}\left(2\left(t - \frac{k}{2}\right)\right). \qquad (4.2.18)$$

This agrees with the formula (3.1.3) for $\phi_S(t)$ in Theorem 3.1 (or the sampling theorem) with $a = \frac{1}{2}$ and $f(t) = \phi_S(t)$.

In Tables 4.1 and 4.2, we give the values of the two-scale sequences $p_{k;N} := p_k$ of the Meyer scaling functions $\phi_{M;N}(t)$ for $N = 0$ and 1, respectively. In Figures 4.7 and 4.8, we show the Meyer scaling functions $\phi_{M;N}(t)$ and wavelets $\psi_{M;N}(t)$ of order $N = 0$ and 1, respectively. The filter characteristics $|\hat{\psi}_{M;N}(\omega)|$ of the Meyer wavelets for $N = 0, 1$ are shown in Figure 4.9.

TABLE 4.1.   *Two-scale sequence $p_k = \phi_{M;0}(\frac{k}{2})$ of the zeroth-order Meyer scaling function.*

k	$p_k = p_{-k}$	k	$p_k = p_{-k}$	k	$p_k = p_{-k}$
0	1.091080	10	$6.1535686E - 03$	20	$1.0445062E - 03$
1	0.6104201	11	$4.9704164E - 03$	21	$-2.1764338E - 03$
2	$-8.1588656E - 02$	12	$-6.7367409E - 03$	22	$1.1081193E - 03$
3	-0.1414711	13	$2.2911504E - 03$	23	$1.0086801E - 03$
4	$5.7279110E - 02$	14	$2.0070113E - 03$	24	$-1.6644868E - 03$
5	$3.1892851E - 02$	15	$-4.2869365E - 03$	25	$6.8705715E - 04$
6	$-2.8294232E - 02$	16	$2.1871887E - 03$	26	$6.3804537E - 04$
7	$6.4225122E - 03$	17	$1.9195266E - 03$	27	$-1.3139109E - 03$
8	$5.2211210E - 03$	18	$-2.9679628E - 03$	28	$6.6719390E - 04$
9	$-1.2126046E - 02$	19	$1.1489112E - 03$	29	$6.2010996E - 04$

TABLE 4.2.   *Two-scale sequence $p_k = \phi_{M;1}(\frac{k}{2})$ of the first-order Meyer scaling function.*

k	$p_k = p_{-k}$	k	$p_k = p_{-k}$	k	$p_k = p_{-k}$
0	1.063513	10	$-8.9132525E - 03$	20	$7.2085112E - 04$
1	0.6237929	11	$-3.5390407E - 03$	21	$6.2655780E - 04$
2	$-5.9431911E - 02$	12	$2.5690419E - 03$	22	$-5.1632337E - 04$
3	-0.1762972	13	$1.8225983E - 03$	23	$-2.1740235E - 04$
4	$4.8477814E - 02$	14	$-1.2258813E - 04$	24	$1.4675227E - 04$
5	$7.5118452E - 02$	15	$-1.9002473E - 03$	25	$1.6278215E - 04$
6	$-3.3952840E - 02$	16	$-3.6127865E - 05$	26	$7.8370795E - 05$
7	$-3.1101495E - 02$	17	$1.8513836E - 03$	27	$-2.8172592E - 04$
8	$1.9765947E - 02$	18	$-4.7928252E - 04$	28	$-6.8847090E - 05$
9	$1.1090688E - 02$	19	$-1.3039149E - 03$	29	$3.5190955E - 04$

$\phi_{M;0}(t)$ $\phi_{M;1}(t)$

FIG. 4.7.   *The zeroth- and first-order Meyer scaling functions.*

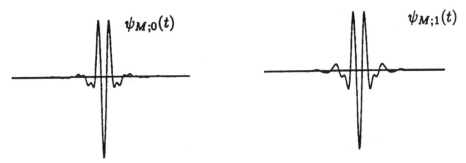

$\psi_{M;0}(t)$ $\psi_{M;1}(t)$

FIG. 4.8.   *The zeroth- and first-order Meyer wavelets.*

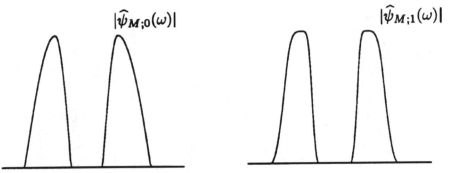

$|\widehat{\psi}_{M;0}(\omega)|$ $|\widehat{\psi}_{M;1}(\omega)|$

FIG. 4.9.   *The bandpass filter characteristics of the Meyer wavelets of order 0 and 1.*

4.3. Spline wavelets of Battle–Lemarié and Strömberg.

The scaling functions $N_1(t), \phi_S(t)$, and $\phi_{M;N}$ for the construction of the Haar, Shannon, and Meyer wavelets, respectively, are already orthonormal, and, hence, orthonormalization is not necessary. When the cardinal B-splines $N_m(t)$, with $m \geq 2$, are used as scaling functions, however, since their integer translates are no longer orthogonal, we need to apply the orthonormalization process in (4.1.15) in order to follow the procedure stated in Theorem 4.2.

For this purpose, we simplify the denominator in the expression (4.1.15) as follows. First observe that the *autocorrelation* of $N_m(t)$, defined by the convolution of $N_m(t)$ with the complex conjugate of its reflection $\overline{N_m(-t)}$, is nothing but the centered $(2m)$th-order cardinal B-spline $N_{2m}(m+t)$ centered at the origin; namely,

$$\int_{-\infty}^{\infty} N_m(x)\overline{N_m(x-t)}\,dx = N_{2m}(m+t). \qquad (4.3.1)$$

On the other hand, it also follows from the Parseval identity (2.2.9b) that

$$\int_{-\infty}^{\infty} N_m(x)\overline{N_m(x-t)}\,dx = \frac{1}{2\pi}\int_{-\infty}^{\infty} |\widehat{N}_m(\omega)|^2 e^{j\omega t}d\omega \qquad (4.3.2)$$

$$= \frac{1}{2\pi}\sum_{k=-\infty}^{\infty}\int_{2\pi k}^{2\pi(k+1)} |\widehat{N}_m(\omega)|^2 e^{j\omega t}d\omega$$

$$= \frac{1}{2\pi}\int_{0}^{2\pi}\left\{\sum_{k=-\infty}^{\infty} |\widehat{N}_m(\omega+2\pi k)|^2\right\} e^{j\omega t}d\omega$$

$$= \frac{1}{2\pi}\int_{0}^{2\pi}\left\{\sum_{k=-\infty}^{\infty} |\widehat{N}_m(\omega+2\pi k)|^2\right\} e^{-j\omega t}d\omega.$$

Hence, equating (4.3.1) with (4.3.2) at $t = n \in \mathbb{Z}$, we see that the nth Fourier coefficient of the 2π-periodic function $\sum_k |\widehat{N}_m(\omega+2\pi k)|^2$ is given by $N_{2m}(m+n)$. That is, we have

$$\sum_{k=-\infty}^{\infty} |\widehat{N}_m(\omega+2\pi k)|^2 = \sum_{n=-\infty}^{\infty} N_{2m}(m+n)e^{jn\omega}$$

$$= \sum_{n=-m+1}^{m-1} N_{2m}(m+n)e^{-jn\omega}. \qquad (4.3.3)$$

Here, we used the support property of $N_{2m}(t)$ in Theorem 3.3 and the fact that the function in (4.3.3) is real valued. In other words, in terms of the *Euler–Frobenius Laurent polynomial* (or E–F L-polynomial)

$$E_m(z) := \sum_{k=-m+1}^{m-1} N_{2m}(m+k)z^k, \qquad (4.3.4)$$

the orthonormalization process (4.1.15) yields

$$\widehat{N}_m^{\perp}(\omega) = \frac{\widehat{N}_m(\omega)}{\{E_m(e^{-j\omega})\}^{1/2}}. \tag{4.3.5}$$

Hence, in view of (3.4.8), the two-scale symbol of the orthonormal scaling function $N_m^{\perp}(t)$ is given by

$$\frac{1}{2}\sum_k p_k e^{-jk\omega/2} = \frac{\widehat{N}_m^{\perp}(\omega)}{\widehat{N}_m^{\perp}(\frac{\omega}{2})} \tag{4.3.6}$$

$$= \frac{\widehat{N}_m(\omega)}{\widehat{N}_m(\frac{\omega}{2})} \left\{ \frac{E_m(e^{-j\omega/2})}{E_m(e^{-j\omega})} \right\}^{1/2}$$

$$= \left(\frac{1 + e^{-j\omega/2}}{2} \right)^m \left\{ \frac{E_m(e^{-j\omega/2})}{E_m(e^{-j\omega})} \right\}^{1/2}.$$

In order to compute the two-scale sequence $\{p_k\}$ as described by (4.3.6), we need the Laurent expansion of the reciprocal of the E–F L-polynomial $E_m(z)$. This can be done in terms of *cardinal spline interpolation* as follows.

Let $L_{2m}(t)$ be the $(2m)$th-order *fundamental cardinal interpolating spline* defined by

$$\begin{cases} L_{2m}(t) = \displaystyle\sum_{k=-\infty}^{\infty} c_k N_{2m}(t + m - k), \\[2mm] L_{2m}(n) = \delta_{n,0}, \qquad n \in \mathbb{Z}, \end{cases} \tag{4.3.7}$$

where the coefficient sequence $\{c_k\}$ is uniquely determined by the *fundamental interpolating* condition $L_{2m}(n) = \delta_{n,0}$, $n \in \mathbb{Z}$, in (4.3.7). In fact, by considering $t = n \in \mathbb{Z}$ in (4.3.7), the *symbol*

$$C_m(z) := \sum_{k=-\infty}^{\infty} c_k z^k \tag{4.3.8}$$

of the sequence $\{c_k\}$ is related to the E–F L-polynomial $E_m(z)$ by

$$C_m(z) = \frac{1}{E_m(z)}. \tag{4.3.9}$$

Hence, by setting

$$z := e^{-j\omega/2} \tag{4.3.10}$$

as usual, in (4.3.6), we have

$$\sum_k p_k z^k = 2 \left(\frac{1+z}{2} \right)^m \{E_m(z)C_m(z^2)\}^{1/2}. \tag{4.3.11}$$

Example 4.4 (the Battle–Lemarié wavelets).

Let $p_k = p_{m;k}$ be the (real and even, i.e., $p_{-k} = p_k$) sequence determined by (4.3.11) with $E_m(z)$ as defined in (4.3.4) and $C_m(z)$ in (4.3.7)–(4.3.8). Then the mth-order Battle–Lemarié wavelet $\psi_{BL;m}(t)$ is given by

$$\psi_{BL;m}(t) = \sum_k (-1)^k p_{1-k} N_m^{\perp}(2t - k). \qquad (4.3.12)$$

Since $\psi_{BL;m}(t)$ is an orthonormal wavelet, the decomposition relation of $N_m^{\perp}(t)$ and $\psi_{BL;m}(t)$ as given in (4.1.19) is valid by using the same sequence $\{p_k\}$ in (4.3.11).

To compute the two-scale sequence $\{p_k\} = \{p_{m;k}\}$, we rely on formula (4.3.11). For instance, for the second-order (i.e., linear) Battle–Lemarié scaling (spline) function $N_2^{\perp}(t)$, we have $m = 2$ and

$$E_2(z) = \sum_{k=-1}^{1} N_4(2 + k) z^k = \frac{1}{6}(z + 4 + z^{-1}). \qquad (4.3.13)$$

Hence, by setting

$$z_0 := -2 + \sqrt{3} , \qquad (4.3.14)$$

it follows from (4.3.9) that

$$C_2(z^2) = \frac{6}{z^2 + 4 + z^{-2}} = \frac{6}{\left(1 - \frac{z_0}{z^2}\right)\left(z^2 - \frac{1}{z_0}\right)} \qquad (4.3.15)$$

$$= -6z_0 \left(\sum_{k=0}^{\infty} z_0^k z^{-2k}\right)\left(\sum_{k=0}^{\infty} z_0^k z^{2k}\right)$$

and with

$$z_1 := -6z_0 \sum_{k=0}^{\infty} z_0^{2k} = -\frac{6z_0}{1 - z_0^2}$$

$$= \frac{6(2 - \sqrt{3})}{4\sqrt{3} - 6} = \sqrt{3},$$

we have

$$C_2(z^2) = z_1 \sum_{k=-\infty}^{\infty} z_0^{|k|} z^{2k} = \sqrt{3} \sum_{k=-\infty}^{\infty} z_0^{|k|} z^{2k},$$

so that

$$E_2(z)C_2(z^2) = \frac{\sqrt{3}}{6}(z + 4 + z^{-1}) \sum_{k=-\infty}^{\infty} z_0^{|k|} z^{2k} \qquad (4.3.16)$$

$$= \frac{\sqrt{3}}{6}\left\{\sum_{k=-\infty}^{\infty} (4z_0^{|k|}) z^{2k} + \sum_{k=-\infty}^{\infty} (z_0^{|k+1|} + z_0^{|k|}) z^{2k+1}\right\},$$

and this yields

$$\{E_2(z)C_2(z^2)\}^{1/2} = 3^{-1/4}\sqrt{2}\left\{1 + \sum_{k \neq 0} \alpha_k z^k\right\}^{1/2}, \qquad (4.3.17)$$

where

$$\begin{cases} \alpha_{2k} = z_0^{|k|}, \\[2mm] \alpha_{2k+1} = \dfrac{z_0^{|k+1|} + z_0^{|k|}}{4}. \end{cases}$$

Finally, by applying (4.3.11) and (4.3.17), we obtain

$$\sum_{k=-\infty}^{\infty} p_k z^k = \frac{(1+z)^2}{\sqrt{2\sqrt{3}}} \left\{ 1 + \frac{1}{2}\sum_{k \neq 0} \alpha_k z^k \right.$$
$$\left. + \sum_{n=2}^{\infty} \frac{(-1)^{n+1}}{2n} \frac{1}{2}\frac{3}{4} \cdots \left(\frac{2n-3}{2n-2}\right) \left(\sum_{k \neq 0} \alpha_k z^k\right)^n \right\},$$

and the values of p_k can be computed by equating coefficients of the same integer powers of z. These values, up to $k = 29$, are listed in Table 4.3.

However, to graph $\psi_{BL;m}(t)$ it is perhaps easier to consider its Fourier transform

$$\widehat{\psi}_{BL;m}(\omega) = \left(\frac{1}{2}\sum_k (-1)^k p_{1-k} e^{-jk\omega/2}\right) \widehat{N}_m^{\perp}\left(\frac{\omega}{2}\right) \qquad (4.3.18)$$

$$= \left(-\frac{1}{2} e^{-j\omega/2} \sum_n (-1)^n p_n e^{jn\omega/2}\right) \frac{\widehat{N}_m(\frac{\omega}{2})}{\{E_m(e^{-j\omega/2})\}^{1/2}}$$

$$= -z\,\frac{1}{2}\sum_k p_k(-z)^k \frac{1}{\{E_m(z)\}^{1/2}} \widehat{N}_m\left(\frac{\omega}{2}\right)$$

$$= -z\left(\frac{1-\bar{z}}{2}\right)^m \left\{\frac{E_m(-z)C_m(z^2)}{E_m(z)}\right\}^{1/2} \left(\frac{1-e^{-j\omega/2}}{j\frac{\omega}{2}}\right)^m$$

$$= -\left(\frac{4}{j\omega}\right)^m \left(\sin^{2m}\left(\frac{\omega}{4}\right)\right) z \left\{\frac{E_m(-z)}{E_m(z)E_m(z^2)}\right\}^{1/2},$$

where the notation $z = e^{-j\omega/2}$ in (4.3.10) is used.

TABLE 4.3.   *Two-scale sequence p_k of the linear Battle–Lemarié scaling function $N_2^{\perp}(t)$.*

k	$p_k = p_{2-k}$	k	$p_k = p_{2-k}$	k	$p_k = p_{2-k}$
1	1.156326630446	13	0.000078285665	25	0.000000020936
2	0.561862928588	14	0.000042442226	26	0.000000011098
3	−.097723548480	15	−.000019542734	27	−.000000005399
4	−.073461813355	16	−.000010527907	28	−.000000002857
5	0.024000684392	17	0.000004921179	29	0.000000001396
6	0.014128834691	18	0.000002638370	30	0.000000000737
7	−.005491761583	19	−.000001247702	31	−.000000000362
8	−.003114029015	20	−.000000666410	32	−.000000000191
9	0.001305843626	21	0.000000318076	33	0.000000000094
10	0.000723562513	22	0.000000169373	34	0.000000000050
11	−.000317202856	23	−.000000081452	35	−.000000000024
12	−.000173504636	24	−.000000043265	36	−.000000000013

For example, it follows from (4.3.13) that the Fourier transform of the second-order (i.e., linear) Battle–Lemarié wavelet is given by

$$\hat{\psi}_{BL;2}(\omega) = \frac{16\sqrt{3}}{\omega^2} \left(\sin^4 \frac{\omega}{4}\right) e^{-j\omega/2} \left\{ \frac{2 - \cos \frac{\omega}{2}}{\left(2 + \cos \frac{\omega}{2}\right)(2 + \cos \omega)} \right\}^{1/2}. \qquad (4.3.19)$$

The graphs of $\psi_{BL;2}(t)$ and $|\hat{\psi}_{BL;2}(\omega)|$ are shown in Figure 4.10.

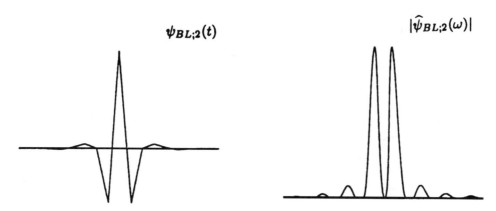

FIG. 4.10.   *The linear Battle–Lemarié wavelet and its filter characteristic.*

Example 4.5 (the Strömberg wavelets).

The Strömberg wavelets are also spline functions but with a certain nonuniform knot sequence. To discuss these wavelets, we first recall from (3.4.18) the notion of mth-order cardinal splines with knot sequence $a\mathbb{Z}$, where $a > 0$ is fixed. More generally, let us consider a (simple) knot sequence

$$\mathbf{t}: \cdots < t_{-1} < t_0 < t_1 < \cdots, \qquad (4.3.20)$$

with $t_n \to +\infty$ and $t_{-n} \to -\infty$ as $n \to +\infty$. Then as a generalization of (4.3.7), the $(2m)$th-order fundamental interpolating splines $L_{\mathbf{t};2m,k}(t)$ with knot sequence $\mathbf{t}$ are defined by

$$\begin{cases} L_{\mathbf{t};2m,k}(t) \in C^{2m-2}, \quad L_{\mathbf{t};2m,k}\big|_{[t_i,t_{i+1}]} \in \pi_{2m-1}, \quad i \in \mathbb{Z}, \\ L_{\mathbf{t};2m,k}(t_\ell) = \delta_{k,\ell} \quad \text{all} \quad k, \ell \in \mathbb{Z}. \end{cases} \qquad (4.3.21)$$

Let us consider the knot sequence $\mathbf{t}$ with

$$t_n = \begin{cases} n & \text{for} \quad n \leq 0, \\ 2n - 1 & \text{for} \quad n \geq 1. \end{cases} \qquad (4.3.22)$$

Then the Strömberg wavelet of order m is the mth-order spline function

$$\psi_{St;m}(t) := c_m L^{(m)}_{\mathbf{t};2m,0}(2t - 1), \qquad (4.3.23)$$

where the knot sequence **t** is given by (4.3.22) and the normalization constant

$$c_m := \frac{1}{\|L_{\mathbf{t};2m,0}^{(m)}\|} \qquad (4.3.24)$$

is chosen so that

$$\|\psi_{St;m}\| = 1. \qquad (4.3.25)$$

In view of (4.3.22), since the mth-order derivative of a $(2m)$th-order spline is an mth-order spline, it follows that $\psi_{St;m}(t)$ is an mth-order spline function with knot sequence

$$\left\{ \ldots, -\frac{3}{2}, -\frac{2}{2}, -\frac{1}{2}, 0, \frac{1}{2}, 1, 2, \ldots \right\}. \qquad (4.3.26)$$

Hence, since this knot sequence is a subsequence of $\frac{1}{2}\mathbb{Z}$, we see that $\psi_{St;m}(t)$ is also an mth-order cardinal spline with knot sequence $\frac{1}{2}\mathbb{Z}$, and we can write

$$\psi_{St;m}(t) = \sum_{k=-\infty}^{\infty} \alpha_k N_m(2t - k). \qquad (4.3.27)$$

From the fundamental interpolating condition of $L_{\mathbf{t};2m,0}(t)$ in (4.3.21) and the knot distribution in (4.3.22), it can be shown that the coefficient sequence $\{\alpha_k\}$ has exponential decay as $|k| \to \infty$. Also, recall from Theorem 3.3(x) that the mth-order cardinal B-spline $N_m(t)$ generates an MRA $\{V_n^m\}$ of L^2. Hence, it follows from (4.3.27) that $\psi_{St;m}(t)$ is in V_1^m. On the other hand, it is also clear from (i), (ii), (iii), and (vii) in Theorem 3.3 that the mth derivative of $N_m(t)$ is a finite linear combination of the delta distributions $\delta(t - k)$; namely,

$$N_m^{(m)}(t) = \sum_{k=0}^{m} (-1)^k \binom{m}{k} \delta(t - k). \qquad (4.3.28)$$

Therefore, from the interpolating property of $L_{\mathbf{t};2m,0}(t)$ in (4.3.21) with $t_0 = 0$ as in (4.3.22), we have, for all integers n, that

$$\begin{aligned}
\langle \psi_{St;m}(t), N_m(t - n) \rangle &= c_m \int_{-\infty}^{\infty} L_{\mathbf{t};2m,0}^{(m)}(2t - 1) N_m(t - n)\,dt \\
&= \frac{(-1)^m}{2^m} c_m \int_{-\infty}^{\infty} L_{\mathbf{t};2m,0} N_m^{(m)}(t - n)\,dt \qquad (4.3.29) \\
&= \frac{(-1)^m}{2^m} c_m \sum_{k=0}^{m} (-1)^k \binom{m}{k} L_{\mathbf{t};2m,0}(2n + 2k - 1) = 0.
\end{aligned}$$

That is, we have

$$\psi_{St;m}(t) \in W_0^m. \qquad (4.3.30)$$

In fact, with a little more care, it can be shown that $\psi_{St;m}(t)$ is orthogonal to $N_m(2t+k)$ for all positive integers k as well. In particular, in view of (4.3.21), (4.3.23), and (4.3.26) we have

$$\langle \psi_{St;m}(t), \psi_{St;m}(t+k) \rangle = 0, \qquad k = 1, 2, \dots . \qquad (4.3.31)$$

It therefore follows from (4.3.25) that

$$\{\psi_{St;m}(t-k): \ k = 0, \pm 1, \dots\}$$

is an orthonormal family. Hence, we have at least partially proved that $\psi_{St;m}(t)$ is an orthonormal wavelet.

However, although the wavelet spaces W_n^m, $n \in \mathbb{Z}$, generated by the orthonormal Strömberg wavelet $\psi_{St;m}(t)$, are orthogonal complementary subspaces of the cardinal spline spaces V_n^m, $n \in \mathbb{Z}$, the wavelet $\psi_{St;m}(t)$ is not constructed by orthonormalization of the scaling function $N_m(t)$. The graphs of $\psi_{St;m}(t)$ for $m = 2, \dots, 5$ are shown in Figure 4.11.

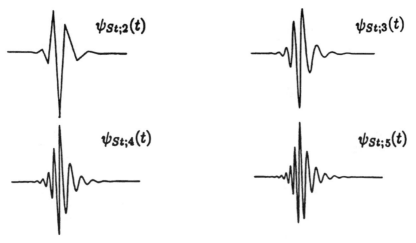

FIG. 4.11. *The Strömberg wavelets of orders $2, 3, 4$, and 5.*

4.4. The Daubechies wavelets.

With the exception of the Haar wavelet introduced in Example 4.1, none of the orthonormal wavelets considered so far have finite support. On the other hand, although the mth-order cardinal B-splines $N_m(t)$, where $m \geq 2$, are compactly supported, they have to be orthonormalized before the procedure given in Theorem 4.2 can be followed. The approach in the following construction of the Daubechies wavelets is to introduce a certain new MRA of L^2 that is generated by compactly supported scaling functions that are already orthonormal, and the two-scale symbols of these scaling functions can be considered as generalizations of the two-scale symbols

$$\left(\frac{1+z}{2}\right)^m$$

of the mth-order B-splines $N_m(t)$ (see (3.4.6)). The generalization is to introduce a polynomial factor $S(z)$, so as to give us some flexibility in achieving
orthonormality, namely, by considering

$$\begin{cases} P(z) = \dfrac{1}{2} \sum_{n=0}^{N} p_n z^n = \left(\dfrac{1+z}{2} \right)^m S(z), \\ S(z) \in \pi_{N-m}, \quad S(1) = 1, \quad \text{and} \quad S(-1) \neq 0 \end{cases} \tag{4.4.1}$$

for some integer $N > m$. The normalization of $S(z)$ in (4.4.1) is to ensure
that $P(1) = 1$ and that the factor $S(z)$ does not increase the multiplicity m
of the root -1 of $P(z)$. Once $S(z)$ has been determined, the orthonormal
Daubechies scaling function $\phi_{D;m}(t)$ can be computed by taking the inverse
Fourier transform of the infinite product

$$\hat{\phi}_{D;m}(\omega) = \prod_{k=1}^{\infty} P(e^{-j\omega/2^k}), \tag{4.4.2}$$

where $P(z)$ is obtained by applying (4.4.1). The two-scale sequence $\{p_k\}$ of
$\phi_{D;m}(t)$ can also be derived from (4.4.1), and the corresponding two-scale
relation

$$\phi_{D;m}(t) = \sum_{k=0}^{N} p_k \phi_{D;m}(2t - k) \tag{4.4.3}$$

gives rise to an alternative (iterative) method to plot the graph of $\phi_{D;m}(t)$,
namely, by taking the limit, as $n \to \infty$, of

$$\begin{cases} \phi_{D;m;n+1}(t) = \sum_{k=0}^{N} p_k \phi_{D;m;n}(2t - k), \\ \text{with} \quad \phi_{D;m;0}(t) = N_2(t), \end{cases} \tag{4.4.4}$$

where $N_2(t)$ is the cardinal linear B-spline. More details in this direction will
be given in section 6.3 of Chapter 6. Of course, the corresponding Daubechies
wavelet is given by

$$\psi_{D;m}(t) = \sum_{k=-N+1}^{1} (-1)^k p_{1-k} \phi_{D;m}(2t - k) \tag{4.4.5}$$

by following (4.1.18). So, the only difficulty is to determine the polynomial
factor $S(z)$ of $P(z)$ in (4.4.1).

For this purpose, we first observe that the Fourier transform representation
of the two-scale relation (4.4.3) is given by

$$\hat{\phi}_{D;m}(\omega) = P(z)\hat{\phi}_{D;m}\left(\frac{\omega}{2}\right), \quad z = e^{-j\omega/2}. \tag{4.4.6}$$

Also, for $\phi_{D;m}(t)$ to be an orthonormal scaling function, it follows from Theorem 4.1 that $\hat{\phi}_{D;m}(\omega)$ must satisfy

$$\sum_{k=-\infty}^{\infty} |\hat{\phi}_{D;m}(\omega + 2\pi k)|^2 = 1 \quad \text{all} \quad \omega. \tag{4.4.7}$$

Hence, by (4.4.7), since

$$\sum_{k=-\infty}^{\infty} |\hat{\phi}_{D;m}(\omega + 2\pi k)|^2 = \sum_{k=-\infty}^{\infty} |P((-1)^k z)|^2 \left|\hat{\phi}_{D;m}\left(\frac{\omega}{2} + \pi k\right)\right|^2 \tag{4.4.8}$$

$$= |P(z)|^2 \sum_{\text{even } k} \left|\hat{\phi}_{D;m}\left(\frac{\omega}{2} + \pi k\right)\right|^2$$

$$+ |P(-z)|^2 \sum_{\text{odd } k} \left|\hat{\phi}_{D;m}\left(\frac{\omega}{2} - \pi + \pi(k+1)\right)\right|^2,$$

we conclude that $P(z)$ must satisfy

$$|P(z)|^2 + |P(-z)|^2 = 1 \quad \text{all} \quad |z| = 1 \tag{4.4.9}$$

for the scaling function $\phi_{D;m}(t)$ to be orthonormal. To transfer the condition (4.4.9) on $P(z)$ to $S(z)$, let us consider the change of variables

$$x := \frac{1 - \cos(\frac{\omega}{2})}{2} = \sin^2\left(\frac{\omega}{4}\right) \tag{4.4.10}$$

and introduce the algebraic polynomial

$$f(x) := |S(e^{-j\omega/2})|^2. \tag{4.4.11}$$

Then we see that (4.4.9) is equivalent to

$$(1 - x)^m f(x) + x^m f(1 - x) = 1 \quad \text{all} \quad x \in \mathbb{R}. \tag{4.4.12}$$

Now, by the binomial theorem, it follows from (4.4.12) that

$$f(x) = (1 - x)^{-m}\{1 - x^m f(1 - x)\} \tag{4.4.13}$$

$$= \sum_{k=0}^{\infty} \binom{m+k-1}{k} x^k \{1 - x^m f(1 - x)\}$$

$$= \sum_{k=0}^{m-1} \binom{m+k-1}{k} x^k + \text{higher-order terms.}$$

Since the polynomial

$$f_0(x) := \sum_{k=0}^{m-1} \binom{m+k-1}{k} x^k \tag{4.4.14}$$

in (4.4.13) already satisfies the identity (4.4.12), and since it is easy to see, by applying the Euclidean algorithm, that there is one and only one polynomial of degree $\leq m-1$ that satisfies (4.4.12), we conclude that $f_0(x)$ is the (unique) polynomial with lowest degree governed by (4.4.12). Going back to the polynomial factor $S(z)$ in (4.4.1), we see that the solution $f(x) = f_0(x)$ in (4.4.14) becomes

$$|S(e^{-j\omega/2})|^2 = \sum_{k=0}^{m-1} \binom{m+k-1}{k} \sin^{2k}\left(\frac{\omega}{4}\right) \tag{4.4.15}$$

simply by putting (4.4.10) into (4.4.14) and then into (4.4.11). Consequently, we must solve for $S(e^{-j\omega/2})$ in (4.4.15). To do so, we need what is called the Riesz lemma.

First, write the expression (4.4.15) as a cosine series; namely,

$$|S(e^{-j\omega/2})|^2 = \frac{a_0}{2} + \sum_{k=1}^{m-1} a_k \cos\left(\frac{k\omega}{2}\right), \tag{4.4.16}$$

where, for each $k = 0, 1, \ldots, m-1$, we have

$$a_k = (-1)^k \sum_{\ell=0}^{m-k-1} \frac{1}{2^{2(k+\ell)-1}} \binom{2(k+\ell)}{\ell} \binom{m+k+\ell-1}{k+\ell}. \tag{4.4.17}$$

Next, consider the polynomial $z^{m-1}|S(z)|^2$. Since it is a reciprocal polynomial with real coefficients, its roots are in reciprocal and complex conjugate pairs. More precisely, we have

$$z^{m-1}|S(z)|^2 = \frac{1}{2} \sum_{\ell=0}^{2m-2} a_{|\ell-m+1|} z^\ell \tag{4.4.18}$$

$$= \frac{a_{m-1}}{2} \left\{ \prod_{k=1}^{K} (z - r_k)\left(z - \frac{1}{r_k}\right) \right\}$$

$$\times \left\{ \prod_{\ell=1}^{L} (z - z_\ell)(z - \bar{z}_\ell)\left(z - \frac{1}{z_\ell}\right)\left(z - \frac{1}{\bar{z}_\ell}\right) \right\},$$

where $K + 2L = m - 1$ and

$$\begin{cases} 0 \neq r_1, \ldots, r_K & \text{real} \\ z_1, \ldots, z_L & \text{complex} (\neq \text{real}). \end{cases} \tag{4.4.19}$$

The Riesz lemma ensures that none of these roots lie on the unit circle. Observe that since

$$\left| \frac{z - z^*}{z - \frac{1}{\bar{z}^*}} \right| = |z^*| \quad \text{all } z = e^{-j\omega/2} \tag{4.4.20}$$

we have some freedom in choosing between r_k and $\frac{1}{r_k}$, $k = 1, \ldots, K$, between z_ℓ and $\frac{1}{z_\ell}$, and between $\bar{z}_\ell$ and $\frac{1}{\bar{z}_\ell}$, $\ell = 1, \ldots, L$ in formulating $S(z)$. What we mean here is that for any such choice, $|S(e^{-j\omega/2})|$ is different from the result of any other choice only by some multiplicative constant that is independent of ω. One possibility is to consider $S(z) = S_{m-1}(z)$, defined by

$$
\begin{cases}
S_{m-1}(z) := \text{constant} \times \displaystyle\prod_{k=1}^{K}(z - r_k) \prod_{\ell=1}^{L}(z - z_\ell)(z - \bar{z}_\ell), \\[2mm]
\text{where } |r_1|, \ldots, |r_K|, |z_1|, \ldots, |z_L| < 1, \\[1mm]
\text{and the constant is chosen so that } S(1) = 1.
\end{cases}
\tag{4.4.21}
$$

This choice of $S_{m-1}(z)$, by selecting the zeros inside the unit circle, corresponds to what is called "minimum-phase digital filtering" with "transfer function"

$$
P_{2m-1}(z) = \frac{1}{2} \sum_{k=0}^{2m-1} p_k z^k := \left(\frac{1+z}{2}\right)^m S_{m-1}(z).
\tag{4.4.22}
$$

In formulating the Daubechies scaling functions $\phi_{D;m}(t)$ in (4.4.2) or (4.4.3) and the Daubechies wavelets $\psi_{D;m}(t)$ in (4.4.5), we will always use the polynomial factors $S_{m-1}(z)$ in (4.4.21). Note that the "filter length" in (4.4.3)–(4.4.5) becomes $2m$, since

$$
N = 2m - 1.
\tag{4.4.23}
$$

It can be shown, however, that for any polynomial factor $S(z)$ in (4.4.1), the corresponding orthonormal scaling functions and wavelets are not symmetric or antisymmetric, unless $S(z) = 1$ and $m = 1$. In addition, the choice of $S(z) = S_{m-1}(z)$ as in (4.4.21) does not necessarily yield scaling functions and wavelets with "least asymmetry." However, we will not pursue this direction any further.

Example 4.6 (the Daubechies wavelet of order 2).

For $m = 2$ in (4.4.17), we have $a_0 = 4$ and $a_1 = -1$, so that, by (4.4.18),

$$
z|S_1(z)|^2 = \frac{1}{2}(-1 + 4z - z^2) = -\frac{1}{2}(z - r_1)\left(z - \frac{1}{r_1}\right),
\tag{4.4.24}
$$

with

$$
r_1 = 2 - \sqrt{3}
\tag{4.4.25}
$$

(and $\frac{1}{r_1} = 2 + \sqrt{3}$). Hence, it follows from (4.4.21) that

$$
S_1(z) = \text{constant} \times (z - r_1)
$$
$$
= \frac{1}{1 - r_1}(z - r_1) = \frac{1}{2}\left\{(\sqrt{3} + 1)z - (\sqrt{3} - 1)\right\},
\tag{4.4.26}
$$

and from (4.4.22) that

$$P_3(z) = \frac{1}{2}\{p_0 + p_1 z + p_2 z^2 + p_3 z^3\} = \left(\frac{1+z}{2}\right)^2 S_1(z) \qquad (4.4.27)$$

$$= \frac{1}{2}\left\{\frac{1-\sqrt{3}}{4} + \frac{3-\sqrt{3}}{4}z + \frac{3+\sqrt{3}}{4}z^2 + \frac{1+\sqrt{3}}{4}z^3\right\}.$$

According to (4.4.2) (or (4.4.3)) and (4.4.5) with $N = 2m - 1 = 3$ as in (4.4.23), the information in (4.4.27) is sufficient to uniquely determine $\phi_{D;2}(t)$ and $\psi_{D;2}(t)$.

In Figures 4.12–4.17, we show the graphs of the Daubechies scaling functions $\phi_{D;m}(t)$ and their corresponding lowpass filter characteristics $|\hat{\phi}_{D;m}(\omega)|$ for $m = 2, \ldots, 7$. The graphs of the corresponding wavelets $\psi_{D;m}(t)$ and their bandpass filter characteristics $|\hat{\psi}_{D;m}(\omega)|$ are shown in Figures 4.18–4.23.

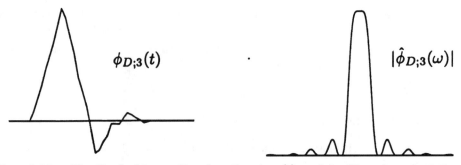

FIG. 4.12.   *The Daubechies scaling function $\phi_{D;2}(t)$ and its filter characteristic.*

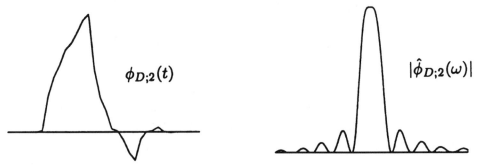

FIG. 4.13.   *The Daubechies scaling function $\phi_{D;3}(t)$ and its filter characteristic.*

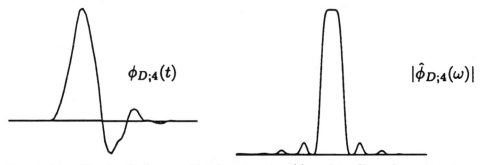

FIG. 4.14.   *The Daubechies scaling function $\phi_{D;4}(t)$ and its filter characteristic.*

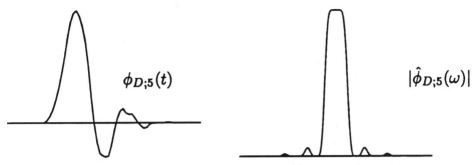

FIG. 4.15. *The Daubechies scaling function $\phi_{D;5}(t)$ and its filter characteristic.*

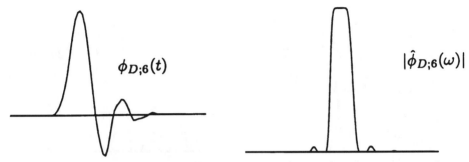

FIG. 4.16. *The Daubechies scaling function $\phi_{D;6}(t)$ and its filter characteristic.*

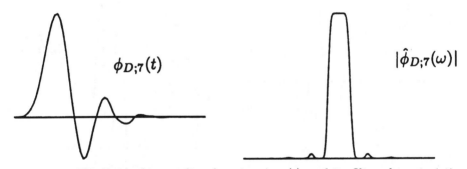

FIG. 4.17. *The Daubechies scaling function $\phi_{D;7}(t)$ and its filter characteristic.*

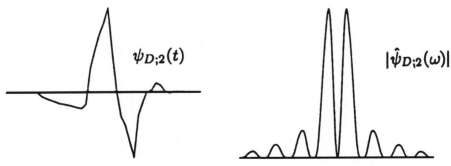

FIG. 4.18. *The Daubechies wavelet $\psi_{D;2}(t)$ and its filter characteristic.*

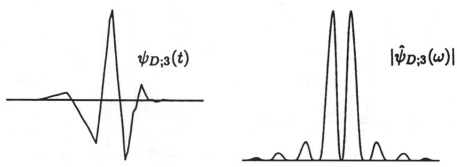

FIG. 4.19.   *The Daubechies wavelet $\psi_{D;3}(t)$ and its filter characteristic.*

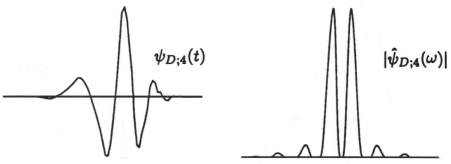

FIG. 4.20.   *The Daubechies wavelet $\psi_{D;4}(t)$ and its filter characteristic.*

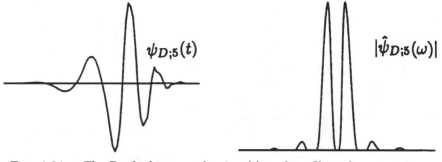

FIG. 4.21.   *The Daubechies wavelet $\psi_{D;5}(t)$ and its filter characteristic.*

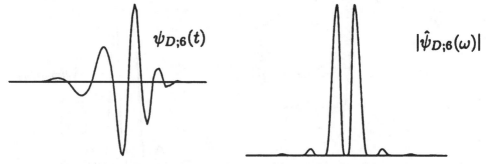

FIG. 4.22.   *The Daubechies wavelet $\psi_{D;6}(t)$ and its filter characteristic.*

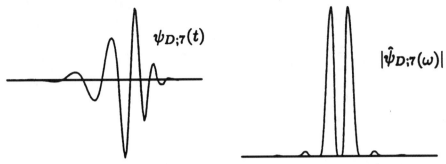

FIG. 4.23. *The Daubechies wavelet $\psi_{D;7}(t)$ and its filter characteristic.*

To study the filter performance, we show both the side-lobe/main-lobe power ratios and the maxima side-lobe/main-lobe ratios (in $-dB$) of $|\hat{\phi}_{D;m}(\omega)|$ and $|\widehat{\psi}_{D;m}(\omega)|$ in Tables 4.4–4.7. In analogy to (3.4.11)–(3.4.12), these ratios are defined as follows:

$$(S/M)(\hat{\phi}_{D;m}) := \frac{\int_{\mathbf{R}\setminus I_m} |\hat{\phi}_{D;m}(\omega)|^2 d\omega}{\int_{I_m} |\hat{\phi}_{D;m}(\omega)|^2 d\omega}, \tag{4.4.28}$$

$$(MS/MM)(\hat{\phi}_{D;m}) := \frac{\max\{|\hat{\phi}_{D;m}(\omega)|\colon \omega \notin I_m\}}{\max\{|\hat{\phi}_{D;m}(\omega)|\colon \omega \in I_m\}}, \tag{4.4.29}$$

$$(S/M)(\widehat{\psi}_{D;m}) := \frac{\int_{(0,\infty)\setminus J_m} |\widehat{\psi}_{D;m}(\omega)|^2 d\omega}{\int_{J_m} |\widehat{\psi}_{D;m}(\omega)|^2 d\omega}, \tag{4.4.30}$$

$$(MS/MM)(\widehat{\psi}_{D;m}) := \frac{\max\{|\widehat{\psi}_{D;m}(\omega)|\colon \omega \notin J_m, \omega > 0\}}{\max\{|\widehat{\psi}_{D;m}(\omega)|\colon \omega \in J_m\}}, \tag{4.4.31}$$

where I_m denotes the support of the main lobe of $|\hat{\phi}_{D;m}(\omega)|$, and J_m denotes the support of the main lobe of $|\widehat{\psi}_{D;m}(\omega)|$ relative to $(0,\infty)$.

TABLE 4.4. *Side-lobe/main-lobe power ratios of $|\hat{\phi}_{D;m}(\omega)|$.*

m	2	3	4	5	6	7
$(S/M)(\hat{\phi}_{D;m})$ in $-dB$	15.960	20.425	24.225	27.687	30.945	34.063

TABLE 4.5. *Maxima side-lobe/main-lobe power ratios of $|\hat{\phi}_{D;m}(\omega)|$.*

m	2	3	4	5	6	7
$(MS/MM)(\hat{\phi}_{D;m})$ in $-dB$	8.037	9.497	10.926	12.319	13.686	15.031

TABLE 4.6. *Side-lobe/main-lobe power ratios of* $|\widehat{\psi}_{D;m}(\omega)|$.

m	2	3	4	5	6	7
$(S/M)(\widehat{\psi}_{D;m})$ in $-dB$	13.520	17.776	21.465	24.860	28.071	31.156

TABLE 4.7. *Maxima side-lobe/main-lobe power ratios of* $|\widehat{\psi}_{D;m}(\omega)|$.

m	2	3	4	5	6	7
$(MS/MM)(\widehat{\psi}_{D;m})$ in $-dB$	7.330	9.054	10.630	12.116	13.545	14.932

Recall from section 1.3 in Chapter 1 that the two-scale sequences

$$\{p_k\} \quad \text{and} \quad \{(-1)^k p_{1-k}\} \tag{4.4.32}$$

that determine an orthonormal scaling function and its corresponding orthonormal wavelet as in (4.1.17)–(4.1.18) are used as filter sequences for signal reconstruction. At the same time, the sequences

$$\left\{\frac{1}{2}p_{-k}\right\} \quad \text{and} \quad \left\{\frac{1}{2}(-1)^k p_{k+1}\right\} \tag{4.4.33}$$

that determine the decomposition relation as in (4.1.19) are used as weight sequences for signal decomposition. Hence, for the orthonormal setting discussed in this chapter, it will be more economical (at least for the purpose of computer storage) to introduce the notation

$$h_k := \frac{1}{\sqrt{2}}p_k \tag{4.4.34}$$

and use $\{h_k\}$ and $\{(-1)^k h_{1-k}\}$ instead of $\{p_k\}$ and $\{(-1)^k p_{1-k}\}$, respectively, as reconstruction sequences. Then, for "perfect reconstruction" the decomposition sequences in (4.4.33) must be multiplied by $\sqrt{2}$. That is, instead of (4.4.33), the decomposition sequences are now $\{h_{-k}\}$ and $\{(-1)^k h_{k-1}\}$. In summary, we have

$$\begin{cases} \text{decomposition sequences: } \{h_{-k}\}, \{(-1)^k h_{k+1}\}; \\ \text{reconstruction sequences: } \{h_k\}, \{(-1)^k h_{1-k}\}. \end{cases} \tag{4.4.35}$$

For the mth-order Daubechies scaling function $\phi_{D;m}(t)$ and wavelet $\psi_{D;m}(t)$, the length of h_k is $2m$. In Table 4.8, we list all the nonzero values of $h_k := h_{m;k}$ for $m = 2, \ldots, 7$ quantized up to 12 decimal places.

TABLE 4.8. *Reconstruction/decomposition sequences for $\phi_{D;m}(t)$ and $\psi_{D;m}(t)$.*

m	n	$h_n = h_{m;n}$	m	n	$h_n = h_{m;n}$	m	n	$h_n = h_{m;n}$
	0	.482962913145		0	.077852054085		12	−.004723204758
2	1	.836516303738	7	1	.396539319482	9	13	−.004281503682
	2	.224143868041		2	.729132090846		14	.001847646883
	3	−.129409522551		3	.469782287405		15	.000230385764
				4	−.143906003929		16	−.000251963189
	0	.332670552950		5	−.224036184994		17	.000039347320
3	1	.806891509311		6	.071309219267			
	2	.459877502118		7	.080612609151			
	3	−.135011020010		8	−.038029936935		0	.026670057901
	4	−.085441273882		9	−.016574541631	10	1	.188176800078
	5	.035226291882		10	.012550998556		2	.527201188932
				11	.000429577973		3	.688459039454
	0	.230377813309		12	−.001801640704		4	.281172343661
4	1	.714846570553		13	.000353713800		5	−.249846424327
	2	.630880767930					6	−.195946274377
	3	−.027983769417		0	.054415842243		7	.127369340336
	4	−.187034811719	8	1	.312871590914		8	.093057364604
	5	.030841381836		2	.675630736297		9	−.071394147166
	6	.032883011667		3	.585354683654		10	−.029457536822
	7	−.010597401785		4	−.015829105256		11	.033212674059
				5	−.284015542962		12	.003606553567
	0	.160102397974		6	.000472484574		13	−.010733175483
5	1	.603829269797		7	.128747426620		14	.001395351747
	2	.724308528438		8	−.017369301002		15	.001992405295
	3	.138428145901		9	−.044088253931		16	−.000685856695
	4	−.242294887066		10	.013981027917		17	−.000116466855
	5	−.032244869585		11	.008746094047		18	.000093588670
	6	.077571493840		12	−.004870352993		19	−.000013264203
	7	−.006241490213		13	−.000391740373			
	8	−.012580751999		14	.000675449406			
	9	.003335725285		15	−.000117476784			
	0	.111540743350		0	.038077947364			
6	1	.494623890398	9	1	.243834674613			
	2	.751133908021		2	.604823123690			
	3	.315250351709		3	.657288078051			
	4	−.226264693965		4	.133197385825			
	5	−.129766867567		5	−.293273783279			
	6	.097501605587		6	−.096840783223			
	7	.027522865530		7	.148540749338			
	8	−.031582039318		8	.030725681479			
	9	.000553842201		9	−.067632829061			
	10	.004777257511		10	.000250947115			
	11	−.001077301085		11	.022361662124			

CHAPTER 5

Biorthogonal Wavelets

Any orthonormal wavelet $\psi(t)$ provides an orthonormal basis

$$\{\psi_{j,k}(t)\colon\ j,k \in \mathbb{Z}\}$$

of the finite-energy space L^2, where the notation of $\psi_{j,k}(t)$ was already introduced in (4.1.20). Hence, any signal $f(t) \in L^2$ has a (generalized Fourier) series representation

$$f(t) = \sum_{j,k} \widehat{d}_{j,k}\psi_{j,k}(t).$$

Recall that the importance of this series expansion is that the (generalized Fourier) coefficients $\widehat{d}_{j,k}$ of $f(t)$ possess very significant time-frequency information of the signal $f(t)$. For each $j, k \in \mathbb{Z}$, the coefficient

$$\widehat{d}_{j,k} = \int_{-\infty}^{\infty} f(t)\overline{\psi_{j,k}(t)}dt = (W_\psi f)\left(\frac{k}{2^j}, \frac{1}{2^j}\right)$$

is the value of the integral wavelet transform (IWT) of $f(t)$, with the orthonormal wavelet $\psi(t)$ itself as the analyzing wavelet, at the time-scale location

$$(b, a) = \left(\frac{k}{2^j}, \frac{1}{2^j}\right).$$

See (2.4.6) for the definition of the IWT and section 2.4 in Chapter 2 for its relevance to time-frequency localization. A plot of these time-scale locations is shown in Figure 5.1.

Observe that although the IWT information of the signal $f(t)$ contained in the coefficient sequence $\{\widehat{d}_{j,k}\}$ is only available on a very sparse set in the time-scale domain as shown in Figure 5.1, this information is sufficient to determine the signal uniquely. In fact, the series representation itself can be used to recover $f(t)$ from its IWT values at $(\frac{k}{2^j}, \frac{1}{2^j})$, $j, k \in \mathbb{Z}$.

Hence, the single function $\psi(t)$, which is an orthonormal wavelet, plays two very important roles in signal analysis. First, it is used as an analyzing wavelet

89

of the IWT to provide certain localized time-frequency information of any finite-energy signal $f(t)$. Second, it is used to give the waveform of the various octaves of the signal. For example, the jth octave $g_j(t)$ of $f(t)$ is given by

$$g_j(t) = \sum_{k=-\infty}^{\infty} \widehat{d}_{j,k} \psi_{j,k}(t) = \sum_{k=-\infty}^{\infty} d_{j,k} \psi(2^j t - k),$$

with $d_{j,k} = 2^{-j/2} \widehat{d}_{j,k}$.

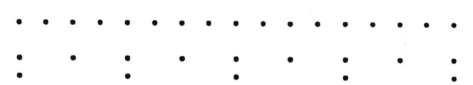

FIG. 5.1.   *Locations in the time-scale domain of the IWT obtained by using a wavelet series representation.*

The objective of this chapter is to study a pair of wavelets $\psi(t)$ and $\widetilde{\psi}(t)$ (called *duals* of each other) that are used to share the workload: one as an analyzing wavelet of the IWT and the other for modeling the octave components of the signal. The fact that their duties can be interchanged will be called the *duality principle*. More precisely, by using the notation introduced in (1.2.3)–(1.2.4), we have, for any finite-energy signal $f(t)$,

$$f(t) = \sum_{j,k} \widehat{d}_{j,k} \psi_{j,k}(t) = \sum_{j,k} \widetilde{\widehat{d}}_{j,k} \widetilde{\psi}_{j,k}(t),$$

where $\widehat{d}_{j,k} = (W_{\widetilde{\psi}} f)\left(\frac{k}{2^j}, \frac{1}{2^j}\right)$ and $\widetilde{\widehat{d}}_{j,k} = (W_\psi f)\left(\frac{k}{2^j}, \frac{1}{2^j}\right)$. We will take advantage of this freedom in the choice of a pair of dual wavelets to construct wavelets with certain desirable filter properties such as asymptotically optimal time-frequency localizations.

5.1. The need for duals.

Let $\{V_n\}$ be a multiresolution analysis (MRA) of L^2 generated by some scaling function $\phi(t)$. As usual, we consider the decomposition

$$V_n = V_{n-1} + W_{n-1}, \qquad n \in \mathbb{Z} \tag{5.1.1}$$

of $\{V_n\}$. However, in general we do not require W_n to be orthogonal to V_n. Instead, we only require

$$V_{n-1} \cap W_{n-1} = \{0\}, \qquad n \in \mathbb{Z}. \tag{5.1.2}$$

Under assumption (5.1.2), the decomposition in (5.1.1) is called a "direct-sum" decomposition. Of course, the orthogonal decomposition with $V_{n-1} \perp W_{n-1}$ is a direct-sum decomposition. So we give ourselves quite a lot of freedom by only assuming (5.1.2). In any case, we will have to assume that the complementary subspaces W_n are generated by a single function $\psi(t)$, called the generating wavelet, and are given by

$$W_n = \text{clos}_{L^2} \langle \psi_{n,k}(t) : k \in \mathbb{Z} \rangle, \tag{5.1.3}$$

where

$$\psi_{n,k}(t) := 2^{n/2} \psi(2^n t - k). \tag{5.1.4}$$

Suppose that $\psi(t)$ is chosen such that it has a dual $\widetilde{\psi}(t)$. By this we mean

$$\langle \psi_{j,k}(t), \widetilde{\psi}_{\ell,m}(t) \rangle = \delta_{j,\ell} \delta_{k,m}, \qquad j, k, \ell, m \in \mathbb{Z}, \tag{5.1.5}$$

where

$$\widetilde{\psi}_{\ell,m}(t) := 2^{\ell/2} \widetilde{\psi}(2^\ell t - m). \tag{5.1.6}$$

Furthermore, we also require the dual wavelet $\widetilde{\psi}(t)$ to be orthogonal to the integer translates of the original scaling function $\phi(t)$. Hence, if we define

$$\widetilde{W}_n = \text{clos}_{L^2} \langle \widetilde{\psi}_{n,k}(t) : k \in \mathbb{Z} \rangle, \tag{5.1.7}$$

then we have

$$\widetilde{W}_n \perp V_n, \quad n \in \mathbb{Z}. \tag{5.1.8}$$

In the construction of $\psi(t)$ and $\widetilde{\psi}(t)$, it is convenient to consider another MRA $\{\widetilde{V}_n\}$ of L^2 that contains $\widetilde{\psi}(t)$ as a generating wavelet. In other words, we need another scaling function $\widetilde{\phi}(t)$ that generates $\{\widetilde{V}_n\}$ in the sense of

$$\widetilde{V}_n = \text{clos}_{L^2} \langle \widetilde{\phi}_{n,k}(t) : k \in \mathbb{Z} \rangle, \qquad n \in \mathbb{Z},$$

with

$$\widetilde{\phi}_{n,k}(t) := 2^{n/2} \widetilde{\phi}(2^n t - k),$$

such that

$$\begin{cases} \widetilde{V}_n = \widetilde{V}_{n-1} + \widetilde{W}_{n-1}, \\ \widetilde{V}_{n-1} \cap \widetilde{W}_{n-1} = \{0\}. \end{cases}$$

Note that in the course of our construction we also require

$$W_n \perp \widetilde{V}_n. \tag{5.1.9}$$

Now, suppose that a finite-energy signal $f(t)$ is represented by some $f_n(t)$ from V_n; that is,

$$f(t) \longmapsto f_n(t) = \sum_k c_{n,k} \phi(2^n t - k). \tag{5.1.10}$$

By repeated applications of the direct-sum decomposition (5.1.1), we have a unique decomposition

$$f_n(t) = g_{n-1}(t) + \cdots + g_{n-M}(t) + f_{n-M}(t), \qquad (5.1.11)$$

where for each $j = n - M, \ldots, n - 1$,

$$g_j(t) = \sum_k d_{j,k} \psi(2^j t - k) \qquad (5.1.12)$$

is in W_j, and

$$f_{n-M}(t) = \sum_k c_{n-M,k} \phi(2^{n-M} t - k) \qquad (5.1.13)$$

is in V_{n-M}. Now, by the duality condition (5.1.5), we see that the coefficients $d_{j,k}$ of the series expansion (5.1.12) of $g_j(t)$ are given by

$$d_{j,k} = 2^j \int_{-\infty}^{\infty} g_j(t) \widetilde{\psi}(2^j t - k) dt, \qquad (5.1.14)$$

and that

$$\int_{-\infty}^{\infty} g_\ell(t) \widetilde{\psi}(2^j t - k) dt = 0, \qquad \ell \neq j \qquad (5.1.15)$$

for all $k \in \mathbb{Z}$. On the other hand, it follows from (5.1.8) and the fact that $V_{n-M} \subset V_j$, $j \geq n - M$ that

$$\int_{-\infty}^{\infty} f_{n-M}(t) \widetilde{\psi}(2^j t - k) dt = 0 \qquad (5.1.16)$$

also. Hence, we may conclude from (5.1.11) and (5.1.14)–(5.1.16) that

$$d_{j,k} = 2^j \int_{-\infty}^{\infty} f_n(t) \widetilde{\psi}(2^j t - k) dt$$

(recall (1.2.12)). That is, the coefficients $d_{j,k}$ are $2^{j/2}$ multiples of the IWT of the analog representation $f_n(t)$ of the original signal $f(t)$ with analyzing wavelet $\widetilde{\psi}(t)$ at the time-scale position $(\frac{k}{2^j}, \frac{1}{2^j})$; namely,

$$d_{j,k} = 2^{j/2} (W_{\widetilde{\psi}} f_n) \left(\frac{k}{2^j}, \frac{1}{2^j} \right) \qquad (5.1.17)$$

(recall (1.2.12)–(1.2.15) and the relation $d_{j,k} = 2^{j/2} \widehat{d}_{j,k}$). As we have seen from (2.4.9)–(2.4.16) in Chapter 2, these values reveal the local time-frequency information of $f_n(t)$ with time-frequency window

$$\left[\frac{k}{2^j} + \frac{t^*}{2^j} - \frac{1}{2^j} \Delta_{\widetilde{\psi}}, \ \frac{k}{2^j} + \frac{t^*}{2^j} + \frac{1}{2^j} \Delta_{\widetilde{\psi}} \right] \qquad (5.1.18)$$

$$\times \left[2^j \omega_+^* - 2^j \Delta_{\underset{\widetilde{\psi}}{\widehat{}}}, \ 2^j \omega_+^* + 2^j \Delta_{\underset{\widetilde{\psi}}{\widehat{}}} \right],$$

where t^* is the center of $\widetilde{\psi}(t)$, and ω_+^* the one-sided center of $\widehat{\widetilde{\psi}}(\omega)$ as defined in (2.4.4). Recall that in the design of the bandpass window function $\widetilde{\psi}(t)$, we require (2.4.14) for the frequency window to slide along the frequency axis. Under this condition, since the frequency window shifts to the right for increasing values of j, the position of the time window (which also narrows for increasing values of j) is important in precisely locating the changes at high frequencies (or large j) of the (nonstationary) signal.

From the above discussions, we see that the wavelet $\psi(t)$ is used to generate the waveform of the octave components $g_j(t)$ of $f_n(t)$, while its dual $\widetilde{\psi}(t)$ is used to analyze the signal $f_n(t)$ itself. We call $\widetilde{\psi}(t)$ an "analyzing wavelet" and $\psi(t)$ a "synthesizing wavelet." Unfortunately, these two wavelets cannot be interchanged, in general, without finding a different signal representation (or model) $f_n(t)$ of the original signal $f(t)$. The reason is that $\widetilde{V}_n$ is different from V_n, in general, and to use $\widetilde{\psi}(t)$ as a synthesizing wavelet, we need to map $f(t)$ to $\widetilde{V}_n$ and use $\widetilde{\phi}(t)$ as the generator instead of $\phi(t)$ as in (5.1.10).

In order to be able to interchange $\psi(t)$ and $\widetilde{\psi}(t)$, we need $\widetilde{V}_n = V_n$, and, hence, as a consequence of (5.1.9), we require

$$W_n \perp V_n.$$

In other words, for the decomposition of signals with finite root mean square (RMS) bandwidths as in (5.1.10)–(5.1.11), the direct-sum decomposition requirement (5.1.1)–(5.1.2) must be the orthogonal decomposition if we wish to interchange $\psi(t)$ and its dual $\widetilde{\psi}(t)$. In this situation, it is most convenient to require $\widetilde{\phi}(t)$ to be the dual scaling function $\phi(t)$ of the MRA $\{V_n\}$ in the sense that

$$\begin{cases} \widetilde{\phi}(t) \in V_0 & \text{and} \\ \langle \phi(t-k), \widetilde{\phi}(t-\ell) \rangle = \delta_{k,\ell}, & k, \ell \in \mathbb{Z}. \end{cases} \tag{5.1.19}$$

From the definition in (5.1.19), it is clear that $\widetilde{\phi}(t)$ is unique. Indeed, if $\eta(t)$ is the difference of two such functions, then

$$\langle \phi(t-k), \eta(t) \rangle = 0, \qquad k \in \mathbb{Z},$$

or $\eta(t)$ is orthogonal to all of V_0, and since $\eta(t)$ is in V_0, it must be the zero function. By a simple modification of the orthonormalization process (4.1.15), we can even write the Fourier transform of this unique dual $\widetilde{\phi}(t)$ of $\phi(t)$ as follows:

$$\widehat{\widetilde{\phi}}(\omega) = \frac{\widehat{\phi}(\omega)}{\sum_k |\widehat{\phi}(\omega + 2\pi k)|^2}. \tag{5.1.20}$$

Observe that this is different from (4.1.15) in that there is no square root in the denominator of (5.1.20).

Observe also that by a straightforward generalization of (4.3.2)–(4.3.3) in the derivation of the Euler–Frobenius Laurent polynomials (E–F L-polynomials) in (4.3.4), the denominator in (5.1.20) can be written as

$$\sum_{k=-\infty}^{\infty} |\widehat{\phi}(\omega + 2\pi k)|^2 = E_\phi(e^{-j\omega}) \tag{5.1.21}$$

with

$$E_\phi(z) := \sum_{n=-\infty}^{\infty} \left\{ \int_{-\infty}^{\infty} \phi(x)\overline{\phi(x-n)}dx \right\} z^n, \qquad |z| = 1. \tag{5.1.22}$$

If $\phi(t)$ has compact support, then $E_\phi(z)$ is a Laurent polynomial, but otherwise $E_\phi(z)$ may be called the Euler–Frobenius Laurent series (or E–F L-series) associated with $\phi(t)$. By (4.1.11), which is equivalent to the Riesz (or stability) condition of a scaling function, we have

$$0 < A \le E_\phi(z) \le B < \infty \qquad \text{all} \quad |z| = 1, \tag{5.1.23}$$

so that $E_\phi(z) \ne 0$ on $|z| = 1$. Consequently, we may write

$$\frac{1}{E_\phi(z)} = \sum_{n=-\infty}^{\infty} a_n z^n, \tag{5.1.24}$$

with $a_n \to 0$ exponentially fast, as $|n| \to \infty$; and (5.1.20)–(5.1.21) together give

$$\widetilde{\phi}(t) = \sum_{n=-\infty}^{\infty} a_n \phi(t-n). \tag{5.1.25}$$

Also, by writing

$$E_\phi(z) = \sum_{n=-\infty}^{\infty} b_n z^n, \tag{5.1.26}$$

we have, by (5.1.20)–(5.1.21),

$$\phi(t) = \sum_{n=-\infty}^{\infty} b_n \widetilde{\phi}(t-n). \tag{5.1.27}$$

Hence, the dual pair of $\phi(t)$ and $\widetilde{\phi}(t)$ in (5.1.19) can be interchanged, and each of them can be expressed in terms of the other by using (5.1.25) and (5.1.27). We will postpone our discussion of change of bases to section 5.3 (see (5.3.33)–(5.3.35)) in this chapter and, more generally, section 6.2 in Chapter 6.

Returning to the signal approximation (or modeling) (5.1.10), since both $\phi(t)$ and $\widetilde{\phi}(t)$ generate the same MRA, we have

$$\begin{aligned}
f_n(t) &= \sum_k c_{n,k} \phi(2^n t - k) \\
&= \sum_k \widetilde{c}_{n,k} \widetilde{\phi}(2^n t - k),
\end{aligned} \tag{5.1.28}$$

so that both sequences $\{c_{n,k}\}$ and $\{\tilde{c}_{n,k}\}$, $k \in \mathbb{Z}$, can be used to represent the same signal $f_n(t)$. Now, in view of the duality property described by (5.1.19), we see that the coefficients $c_{n,k}$ and $\tilde{c}_{n,k}$ of the series representations in (5.1.28) are given by

$$
\begin{cases}
c_{n,k} = 2^n \langle f_n(t), \tilde{\phi}(2^n t - k) \rangle = 2^n \displaystyle\int_{-\infty}^{\infty} f_n(t) \tilde{\phi}(2^n t - k) dt, \\[2mm]
\tilde{c}_{n,k} = 2^n \langle f_n(t), \phi(2^n t - k) \rangle = 2^n \displaystyle\int_{-\infty}^{\infty} f_n(t) \phi(2^n t - k) dt,
\end{cases}
\tag{5.1.29}
$$

respectively. That is, the coefficients in (5.1.28) are filter outputs of the input signal $f_n(t)$, with lowpass filters $2^n \tilde{\phi}(2^n t - k)$ and $2^n \phi(2^n t - k)$. More precisely, if we set

$$
\begin{cases}
h(t) := 2^n \phi(2^n t), \quad \text{so that} \\[2mm]
\Delta_h = 2^{-n} \Delta_\phi \quad \text{and} \quad \Delta_{\hat{h}} = 2^n \Delta_{\hat{\phi}},
\end{cases}
\tag{5.1.30}
$$

then the second formula in (5.1.29) can be written as

$$
\tilde{c}_{n,k} = \int_{-\infty}^{\infty} f_n(t) \overline{h\left(t - \frac{k}{2^n}\right)} dt
\tag{5.1.31}
$$

$$
= \frac{1}{2\pi} \int_{-\infty}^{\infty} \hat{f}_n(\omega) e^{j(k/2^n)\omega} \overline{\hat{h}(\omega)} d\omega.
$$

Hence, by (2.2.10)–(2.2.14), it follows from (5.1.30) that the coefficient $\tilde{c}_{n,k}$ reveals the time-frequency content of the signal $f_n(t)$, with time-frequency window

$$
\left[t^* + \frac{k}{2^n} - 2^{-n} \Delta_\phi, \; t^* + \frac{k}{2^n} + 2^{-n} \Delta_\phi \right] \times \left[-2^n \Delta_{\hat{\phi}}, \; 2^n \Delta_{\hat{\phi}} \right],
\tag{5.1.32}
$$

where t^* is the center of $\phi(2^n t)$. Observe that the subscript k of $\tilde{c}_{n,k}$ can be used to identify the center $t^* + \frac{k}{2^n}$ of the time window in (5.1.32) and that the radius $2^{-n} \Delta_\phi$ of this window is small for large values of n. On the other hand, the frequency range increases with increasing values of n. Consequently, for signals with large RMS bandwidths, the coefficients $c_{n,k}$ or $\tilde{c}_{n,k}$ in the series expansions (5.1.28) of $f_n(t)$ give very precise time location of the signal $f_n(t)$ at which the RMS bandwidth is $2(2^n \Delta_{\tilde{\phi}})$ or $2(2^n \Delta_{\hat{\phi}})$, respectively. The effectiveness of this information depends on the filter characteristics of $\tilde{\phi}(t)$ or $\phi(t)$. One of the important considerations in this regard is the size of the side-lobe/main-lobe ratios, such as those of the cardinal B-splines in Tables 3.1–3.2 and those of the Daubechies scaling functions in Tables 4.4–4.5. Another important consideration is the flatness of the magnitude spectrum $|\hat{\tilde{\phi}}(\omega)|$ or $|\hat{\phi}(\omega)|$ in a neighborhood of $\omega = 0$. (See Figures 3.4–3.7 for cardinal B-splines, Figure 4.5 for the Shannon and Meyer scaling functions, and Figures 4.12–4.17 for the Daubechies scaling functions.) Unfortunately, although a "maximally flat" lowpass filter is usually desirable, it gives poor time localization. In fact, all high-order orthonormal scaling functions imitate the Shannon sampling function $\phi_S(t)$ in (3.1.5) and, hence, have large RMS durations.

More precisely, we have the following result. Although this theorem is concerned with the Daubechies scaling functions, we remark that all orthonormal scaling functions of high orders discussed in Chapter 4 have similar time-frequency localization performance.

Theorem 5.1. Let $\phi_S(t)$ be the Shannon sampling function and $\phi_{D;m}(t)$ the mth-order Daubechies scaling functions. Then as $m \to \infty$,

$$\| \, |\widehat{\phi}_{D;m}(\omega)| - |\widehat{\phi}_S(\omega)| \, \| = \| \, |\widehat{\phi}_{D;m}(\omega)| - \chi_{[-\pi,\pi]}(\omega)\| \to 0 \qquad (5.1.33)$$

and

$$\Delta_{\phi_{D;m}} \Delta_{\widehat{\phi}_{D;m}} \to \infty. \qquad (5.1.34)$$

Nonetheless, since our concern is not the short-time Fourier transform (STFT), time localization in lowpass filtering is really not an important issue. What is important in our study is that by mapping a finite-energy signal $f(t)$ to a signal $f_n(t)$ in an MRA space V_n as shown in (5.1.10), the signal representor $f_n(t)$ of $f(t)$ can be decomposed as (5.1.11) for any desirable positive integer M.

Let us first observe that the RMS bandwidth of the $f_{n-M}(t)$ is $2^{n-M+1}\Delta_{\widehat{\phi}}$, which tends to 0 as $M \to \infty$, so that $f_{n-M}(t)$ represents the "DC" component of the signal representor $f_n(t)$. As for the octave components $g_{n-1}(t), \ldots,$ $g_{n-M}(t)$ of $f_n(t)$ in (5.1.11), we have already seen from (5.1.17) that their coefficient sequences $d_{j,k}$, $j = n - M, \ldots, n - 1$, are $2^{j/2}$ multiples of the IWT of $f_n(t)$ at the positions

$$\left(\frac{k}{2^j}, \frac{1}{2^j} \right), \qquad k \in \mathbb{Z},$$

in the time-scale domain, using $\widetilde{\psi}(t)$ as the analyzing wavelet. Hence, a study of the bandpass filter characteristic of the analyzing wavelet $\widetilde{\psi}(t)$ is important.

Again, since we are interested in interchanging the roles of analysis and synthesis of the wavelet $\psi(t)$ and its dual $\widetilde{\psi}(t)$, we will only consider orthogonal decomposition $V_{j+1} = V_j + W_j$ with $V_j \perp W_j$, so that in view of the nested property of the MRA, we have the orthogonal decomposition in (4.1.2) of the finite-energy space L^2; that is,

$$L^2 = \bigoplus_{n=-\infty}^{\infty} W_n, \qquad W_j \perp W_\ell, \quad j \neq \ell. \qquad (5.1.35)$$

We emphasize that $\{\psi_{j,k}(t)\}$ does not need to be an orthonormal basis to yield an orthogonal decomposition of L^2 as in (5.1.35). In fact, it is clear that, by using the notation in (5.1.4), we have

$$W_j \perp W_\ell \Longleftrightarrow \langle \psi_{j,k}(t), \psi_{\ell,m}(t) \rangle = 0 \quad \text{all} \quad k, m \in \mathbb{Z}, \qquad (5.1.36)$$

for any $j \neq \ell$, and there is no indication in (5.1.36) of orthogonality for $j = \ell$.

Definition 5.1. *A function $\psi(t) \in L^2$ is called a semiorthogonal wavelet if it is stable in the sense that some constants A and B, with $0 < A \leq B < \infty$, exist such that*

$$A \leq \sum_k |\widehat{\psi}(\omega + 2\pi k)|^2 \leq B \quad a.e., \tag{5.1.37}$$

and if the spaces W_j, $j \in \mathbb{Z}$, it generates as in (5.1.3), give an orthogonal decomposition of the finite-energy space L^2 as described by (5.1.35).

Hence, by (5.1.36), any semiorthogonal wavelet $\psi(t)$ satisfies

$$\langle \psi(t), \psi(2^j t - k) \rangle = 0, \quad j, k \in \mathbb{Z}, \quad j \neq 0. \tag{5.1.38}$$

Analogous to (5.1.20), the dual $\widetilde{\psi}(t)$ of a semiorthogonal wavelet is determined by considering its Fourier transform

$$\widehat{\widetilde{\psi}}(\omega) = \frac{\widehat{\psi}(\omega)}{\sum_k |\widehat{\psi}(\omega + 2\pi k)|^2}. \tag{5.1.39}$$

Observe that if $A = B = 1$ in (5.1.37), then $\{\psi(t - k) \colon k \in \mathbb{Z}\}$ is orthonormal (see Theorem 4.1), and, according to (5.1.39), $\widetilde{\psi}(t) = \psi(t)$. That is, an orthonormal wavelet is a semiorthogonal wavelet with its dual equal to itself. In the next section, we will give a characterization of all semiorthogonal wavelets corresponding to a given scaling function $\phi(t)$. When a cardinal B-spline $N_m(t)$, $m \geq 2$ is used as the scaling function $\phi(t)$, we will also formulate, in section 5.2, an explicit expression of the corresponding compactly supported semiorthogonal spline wavelet $\psi_m(t)$ with minimum support. This wavelet, which happens to be symmetric for even m and antisymmetric for any odd integer m, will be called the mth-order *cardinal B-wavelet*.

On the other hand, there is no orthonormal wavelet, except the Haar wavelet, that has compact support and is symmetric or antisymmetric. In the following, we will see that orthonormal wavelets of high order do not give good time-frequency localization either. In this regard, let us first recall the uncertainty principle for bandpass filtering as stated in (2.4.21). We will see in section 5.4 that a class of semiorthogonal wavelets, including the compactly supported cardinal B-wavelets $\psi_m(t)$, have asymptotically optimal time-frequency localization performance.

For orthonormal wavelets, we only discuss the asymptotic behavior of the Daubechies wavelets, although a similar result holds for the other orthonormal wavelets discussed in Chapter 4.

Theorem 5.2. *Let $\psi_S(t)$ be the Shannon wavelet and $\psi_{D;m}(t)$ the Daubechies wavelets of order m. Then as $m \to \infty$,*

$$\| \, |\widehat{\psi}_{D;m}(\omega)| - |\widehat{\psi}_S(\omega)| \, \| \to 0 \tag{5.1.40}$$

and

$$\Delta_{\psi_{D;m}} \Delta^+_{\widehat{\psi}_{D;m}} \to \infty. \tag{5.1.41}$$

Hence, for better time-frequency localization, one should not use very high order orthonormal wavelets.

5.2. Compactly supported spline wavelets.

Let $\phi(t)$ be a compactly supported scaling function that generates some MRA $\{V_j\}$ of L^2. The objective of this section is to describe a method for constructing its corresponding compactly supported and, in fact, minimally supported semiorthogonal wavelet $\psi(t)$. Its dual, the duality principle, and the decomposition and reconstruction sequences will be studied in section 5.3. (For orthonormal scaling functions and orthonormal wavelets, these sequences are given in (4.4.35) by using the definition of h_n in (4.4.34).) We will then apply this method to the mth-order cardinal B-splines $N_m(t)$ and, hence, obtain explicit formulas of the mth-order cardinal B-wavelets $\psi_m(t)$.

By a suitable integer shift, if necessary, we will always assume that the support of the compactly supported scaling function $\phi(t)$ is $[0, M]$ for some positive integer M, so that the two-scale relation of $\phi(t)$ can be written as

$$\phi(t) = \sum_{k=0}^{M} p_k \phi(2t - k), \qquad p_0 p_M \neq 0. \tag{5.2.1}$$

For convenience, we will only consider real-valued functions $\phi(t)$. Hence, the autocorrelation function

$$\Phi(t) := \int_{-\infty}^{\infty} \phi(x - t)\overline{\phi(x)}dx = \int_{-\infty}^{\infty} \phi(x - t)\phi(x)dx \tag{5.2.2}$$

relative to $\phi(t)$ is an even (and, hence, symmetric) function, with support given by

$$\text{supp } \Phi = [-M, M]. \tag{5.2.3}$$

Let n_ϕ be the largest integer for which $\Phi(n_\phi) \neq 0$. Then since $\Phi(t)$ is continuous, (5.2.3) implies that $\Phi(M) = 0$, so that

$$0 \leq n_\phi \leq M - 1, \tag{5.2.4}$$

and it follows from (5.1.21)–(5.1.22) that

$$\begin{cases} \displaystyle\sum_{k=-\infty}^{\infty} |\hat{\phi}(\omega + 2\pi k)|^2 = E_\phi(e^{-j\omega}), \\[2mm] \displaystyle E_\phi(z) := \sum_{k=-n_\phi}^{n_\phi} \Phi(k)z^k, \end{cases} \tag{5.2.5}$$

where $E_\phi(z)$ is the E–F L-polynomial associated with the scaling function $\phi(t)$.

Example 5.1. If $\phi(t)$ is an orthonormal scaling function, then $\Phi(k) = \delta_{k,0}$, so that $n_\phi = 0$ and the E–F L-polynomial is simply $E_\phi(z) = 1$.

Example 5.2. Let $\phi(t)$ be the mth-order B-spline $N_m(t)$. Then $\Phi(t) = N_{2m}(t+m)$, which is the "centered" $(2m)$th-order B-spline with supp $\Phi = [-m, m]$. Hence, $n_\phi = m - 1$ and

$$E_{N_m}(z) = \sum_{k=-m+1}^{m-1} N_{2m}(m+k)z^k, \tag{5.2.6}$$

which agrees with the E–F L-polynomial $E_m(z)$ introduced in (4.3.4).

Let us return to an arbitrary compactly supported scaling function $\phi(t)$ that generates an MRA $\{V_n\}$ of L^2, and consider the orthogonal decomposition

$$V_1 = V_0 + W_0, \qquad W_0 \perp V_0 \tag{5.2.7}$$

as discussed in section 5.1. Before we try to determine the minimally supported functions in W_0 that generate all of W_0, let us first investigate the structure of any function $\psi(t)$ in W_0. In the first place, if $\psi(t) \in W_0$, then since $W_0 \subset V_1$, we may write

$$\psi(t) = \sum_k r_k \phi(2t - k), \qquad \{r_k\} \in \ell^2. \tag{5.2.8}$$

By setting

$$R(z) = \frac{1}{2} \sum_k r_k z^k, \tag{5.2.9}$$

which is a function in $L^2(|z| = 1)$, the space of 2π-periodic signals with finite energy, we see that expression (5.2.8) may be formulated in the frequency domain as

$$\widehat{\psi}(\omega) = R(z)\widehat{\phi}\left(\frac{\omega}{2}\right), \qquad z = e^{-j\omega/2}. \tag{5.2.10}$$

Of course the same transformation can be applied to the two-scale relation (5.2.1) of the scaling function $\phi(t)$ itself, yielding

$$\widehat{\phi}(\omega) = P(z)\widehat{\phi}\left(\frac{\omega}{2}\right), \qquad z = e^{-j\omega/2}, \tag{5.2.11}$$

where

$$P(z) = \frac{1}{2} \sum_{k=0}^{M} p_k z^k$$

is called the *two-scale symbol* of $\phi(t)$. Now, by applying the Parseval identity, (5.2.11) and (5.2.10), the 4π-periodicity of $z = e^{-j\omega/2}$, (5.2.5), and the fact that $e^{-j(\omega - 2\pi)/2} = -e^{-j\omega/2} = -z$, consecutively, we have, for every integer n,

$$\langle \phi(t - n), \psi(t) \rangle := \int_{-\infty}^{\infty} \phi(t - n)\overline{\psi(t)}dt \tag{5.2.12}$$

$$= \frac{1}{2\pi} \int_{-\infty}^{\infty} \widehat{\phi}(\omega)e^{-jn\omega}\overline{\widehat{\psi}(\omega)}d\omega$$

$$= \frac{1}{2\pi} \int_{-\infty}^{\infty} \left| \widehat{\phi} \left(\frac{\omega}{2} \right) \right|^2 P(z) \overline{R(z)} e^{-jn\omega} d\omega$$

$$= \frac{1}{2\pi} \sum_{k=-\infty}^{\infty} \int_{4\pi k}^{4\pi(k+1)} \left| \widehat{\phi} \left(\frac{\omega}{2} \right) \right|^2 P(z) \overline{R(z)} e^{-jn\omega} d\omega$$

$$= \frac{1}{2\pi} \int_{0}^{4\pi} \left\{ \sum_{k=-\infty}^{\infty} \left| \widehat{\phi} \left(\frac{\omega}{2} + 2\pi k \right) \right|^2 \right\} P(z) \overline{R(\omega)} e^{-jn\omega} d\omega$$

$$= \frac{1}{2\pi} \int_{0}^{4\pi} E_\phi(z) P(z) \overline{R(z)} e^{-jn\omega} d\omega$$

$$= \frac{1}{2\pi} \int_{0}^{2\pi} \{ E_\phi(z) P(z) \overline{R(z)} + E_\phi(-z) P(-z) \overline{R(-z)} \} e^{-jn\omega} d\omega.$$

Hence, for $\psi(t)$ to be in W_0 it must be orthogonal to $\phi(t - n)$ for all $n \in \mathbb{Z}$ so that the quantity inside the braces of the last integral in (5.2.12), being a 2π-periodic function in $L^2(|z| = 1)$, must be identically zero. That is, the symbol function $R(z)$ in (5.2.10) must satisfy

$$E_\phi(z) P(z) \overline{R(z)} + E_\phi(-z) P(-z) \overline{R(-z)} = 0, \qquad |z| = 1. \qquad (5.2.13)$$

The general solution of (5.2.13) is given by

$$\begin{cases} R(z) = zK(z^2) P\left(-\frac{1}{z} \right) E_\phi(-z), & |z| = 1 \\ \\ \text{for any } K(z) \in L^2(|z| = 1). \end{cases} \qquad (5.2.14)$$

In the formulation of $R(z)$ in (5.2.14), we have assumed that the coefficients p_k of the polynomial $P(z)$ are real, and we have also used the fact that $\Phi(-k) = \Phi(k)$ implies $E_\phi(z^{-1}) = E_\phi(z)$ for $|z| = 1$.

Example 5.3. As a continuation of Example 5.1, let us consider any orthonormal scaling function $\phi(t)$ and choose $K(z) = -1$. Then since $E_\phi(z) = 1$, we have

$$R(z) = -zP\left(-\frac{1}{z} \right) = \frac{1}{2} \sum_k -z p_k \left(-\frac{1}{z} \right)^k \qquad (5.2.15)$$

$$= \frac{1}{2} \sum_k p_k (-z)^{-k+1} = \frac{1}{2} \sum_k (-1)^k p_{1-k} z^k,$$

so that, by (5.2.9),

$$r_k = (-1)^k p_{1-k}. \qquad (5.2.16)$$

Hence, it follows from (5.2.8), that

$$\psi(t) = \sum_k (-1)^k p_{1-k} \phi(2t - k).$$ (5.2.17)

This is the two-scale relation of an orthonormal wavelet as formulated in (4.1.18) in terms of its corresponding orthonormal scaling function $\phi^{\perp}(t) = \phi(t)$.

In general, for any compactly supported scaling function $\phi(t)$, since both $P(z)$ and $E_\phi(z)$ are Laurent polynomials, it follows from (5.2.14) that $R(z)$ is again a Laurent polynomial provided that $K(z)$ is also a Laurent polynomial. In fact, the support (or length) of the coefficient sequence of $R(z)$ is the smallest (or shortest) if $K(z)$ is a monomial. That is, we have the following result.

Theorem 5.3. *Let $\phi(t)$ be a compactly supported scaling function with two-scale relation (5.2.1). Then the corresponding compactly supported semiorthogonal wavelets $\psi(t)$ with minimum support are governed by the two-scale relation*

$$\psi(t) = \sum_k q_k \phi(2t - k)$$ (5.2.18)

or, equivalently,

$$\widehat{\psi}(\omega) = Q(z)\widehat{\phi}\left(\frac{\omega}{2}\right), \qquad z = e^{-j\omega/2},$$ (5.2.19)

with

$$Q(z) := \frac{1}{2}\sum_k q_k z^k = cz^{2N+1} P\left(-\frac{1}{z}\right) E_\phi(-z),$$ (5.2.20)

where N is any integer and c any nonzero constant. In particular, by considering the largest integer

$$N := \left\lfloor \frac{M + n_\phi}{2} \right\rfloor$$ (5.2.21)

not exceeding $(M + n_\phi)/2$, $Q(z)$ becomes an algebraic polynomial in z and (5.2.18) can be written more precisely as

$$\psi(t) = \sum_{k=0}^{M+2n_\phi} q_k \phi(2t - k).$$ (5.2.22)

In formulating (5.2.20), we have chosen $K(z)$ in (5.2.14) to be cz^N. Hence, by (5.2.5) the choice of N in (5.2.21) yields (5.2.22).

Example 5.4. For the mth-order cardinal B-splines $N_m(t)$, $m \geq 1$, we have $M = m$ and $n_\phi = m - 1$, so that the integer N in (5.2.21) becomes $m - 1$.

Hence, setting $c = -1$ in (5.2.20) and applying (2.4.6) and (5.2.6), we have

$$Q(z) = -z^{2m-1} P\left(-\frac{1}{z}\right) E(-z) \tag{5.2.23}$$

$$= -z^{2m-1} \left(\frac{z-1}{2z}\right)^m \sum_{n=-m+1}^{m-1} (-1)^n N_{2m}(m+n) z^n$$

$$= \left(\frac{1-z}{2}\right)^m \sum_{k=0}^{2m-2} (-1)^k N_{2m}(k+1) z^k = \frac{1}{2} \sum_{k=0}^{3m-2} q_k z^k,$$

with

$$q_k = \frac{(-1)^k}{2^{m-1}} \sum_{\ell=0}^{m} \binom{m}{\ell} N_{2m}(k-\ell+1), \qquad k = 0, \ldots, 3m-2. \tag{5.2.24}$$

So, for each positive integer m, the mth-order cardinal B-wavelet (or cardinal spline B-wavelet) is given by

$$\psi_m(t) = \sum_{k=0}^{3m-2} q_k N_m(2t - k). \tag{5.2.25}$$

In addition, by (2.4.8), (5.2.19), and (5.2.23), the bandpass filter characteristic of $\psi_m(t)$ can be written as

$$|\hat{\psi}_m(\omega)| = |\omega|^{-m} \left(2\sin\frac{\omega}{4}\right)^{2m} \left\{ N_{2m}(m) + 2\sum_{k=1}^{m-1} (-1)^k N_{2m}(m+k) \cos\frac{k\omega}{2} \right\}, \tag{5.2.26}$$

where the values of $N_{2m}(m+k)$ can be computed recursively by applying (viii) in Theorem 3.3; namely,

$$\begin{cases} N_n(k) = \dfrac{k}{n-1} N_{n-1}(k) + \dfrac{n-k}{n-1} N_{n-1}(k-1), \\[2mm] \text{with } N_2(k) = \delta_{k,1}, \qquad k \in \mathbb{Z}, \end{cases} \tag{5.2.27}$$

where $n = 3, \ldots, 2m$.

The graphs of the cardinal B-wavelets $\psi_m(t)$ along with their bandpass filter characteristics for $m = 2, \ldots, 6$ (i.e., from linear to quintic piecewise polynomials) are shown in Figures 5.2–5.6.

Observe that each B-wavelet $\psi_m(t)$ has compact support (or finite time duration) with

$$\text{supp } \psi_m = [0, 2m - 1], \tag{5.2.28}$$

and that $\psi_m(t)$ is symmetric for even m and antisymmetric for odd m with respect to its center

$$t_m^* := \frac{2m-1}{2}. \tag{5.2.29}$$

FIG. 5.2.   *Linear B-wavelet* $\psi_2(t)$ *and its filter characteristic.*

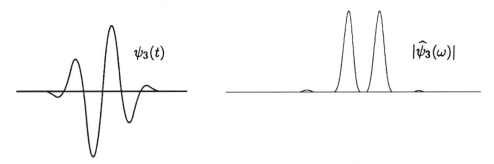

FIG. 5.3.   *Quadratic B-wavelet* $\psi_3(t)$ *and its filter characteristic.*

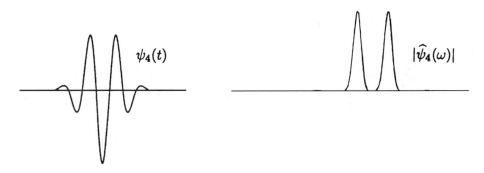

FIG. 5.4.   *Cubic B-wavelet* $\psi_4(t)$ *and its filter characteristic.*

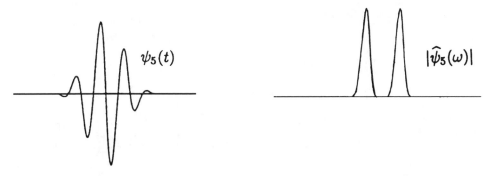

FIG. 5.5.   *Quartic B-wavelet* $\psi_5(t)$ *and its filter characteristic.*

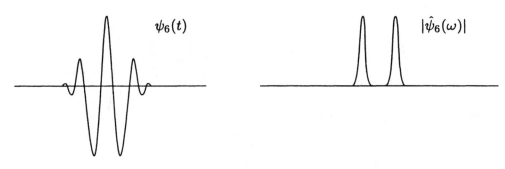

FIG. 5.6.   *Quintic B-wavelet $\psi_6(t)$ and its filter characteristic.*

Hence, bandpass filtering by $\psi_m(t)$ is linear phase or at least generalized linear phase. To study the filter performance, we give the values (in $-dB$) of the side-lobe/main-lobe power ratios and maxima side-lobe/main-lobe ratios $(S/M)(\widehat{\psi}_m)$ and $(MS/MM)(\widehat{\psi}_m)$, as defined in (4.4.30)–(4.4.31), in Tables 5.1 and 5.2, respectively.

TABLE 5.1.   *Side-lobe/main-lobe power ratios of $|\widehat{\psi}_m(\omega)|$.*

	2	3	4	5	6
$(S/M)(\widehat{\psi}_m)$ in $-dB$	20.13775	30.88587	41.29142	51.59685	61.88989

TABLE 5.2.   *Maxima side-lobe/main-lobe power ratios of $|\widehat{\psi}_m(\omega)|$.*

m	2	3	4	5	6
$(MS/MM)(\widehat{\psi}_m)$ in $-dB$	10.256437	15.383628	20.511496	25.639370	32.907667

Finally, to determine the size of the time-frequency window of $\psi_m(t)$, we need its one-sided center

$$\omega_{m,+}^* := \frac{\int_0^\infty \omega |\widehat{\psi}_m(\omega)|^2 d\omega}{\int_0^\infty |\widehat{\psi}_m(\omega)|^2 d\omega} \tag{5.2.30}$$

as introduced in (2.4.4). These values for $m = 2, \ldots, 6$ are shown in Table 5.3, and the values of $\Delta_{\psi_m} \Delta_{\widehat{\psi}_m}^+$ are shown in Table 5.4. Recall from (2.4.20) that the uncertainty lower bound is $\frac{1}{2}$.

TABLE 5.3.   *One-sided centers of $|\widehat{\psi}_m(\omega)|$.*

m	2	3	4	5	6
$\omega_{m,+}^*$	5.331705	5.181498	5.164206	5.158902	5.155909

TABLE 5.4. *Values of* $\Delta_{\psi_m}\Delta_{\widehat{\psi}_m}^+$ *of the B-wavelets.*

m	$\Delta_{\psi_m}\Delta_{\widehat{\psi}_m}^+$	m	$\Delta_{\psi_m}\Delta_{\widehat{\psi}_m}^+$
2	0.9196784	7	0.5002750
3	0.5385703	8	0.5002200
4	0.5047686	9	0.5001843
5	0.5009224	10	0.5001580
6	0.5003949	11	0.5001377

5.3. The duality principle.

If $\psi(t)$ and $\widetilde{\psi}(t)$ are dual wavelets in the sense of (5.1.5), then every finite-energy signal $f(t) \in L^2$ has two series representations:

$$f(t) = \sum_{j,k} \widehat{d}_{j,k}\psi_{j,k}(t) = \sum_{j,k} \widetilde{d}_{j,k}\widetilde{\psi}_{j,k}(t), \qquad (5.3.1)$$

where the coefficient sequences are the IWT values

$$\begin{cases} \widehat{d}_{j,k} = (W_{\widetilde{\psi}}f)\left(\dfrac{k}{2^j}, \dfrac{1}{2^j}\right), \\[2mm] \widetilde{d}_{j,k} = (W_{\psi}f)\left(\dfrac{k}{2^j}, \dfrac{1}{2^j}\right), \qquad j,k \in \mathbb{Z}. \end{cases} \qquad (5.3.2)$$

Hence, the roles of $\psi(t)$ and $\widetilde{\psi}(t)$, as a local basis generating function and as an analyzing wavelet for the IWT, can be interchanged. However, in applications to signal analysis, $f(t)$ must first be represented by an RMS bandlimited signal $f_n(t)$, say, as shown in (5.1.10). Then writing

$$f_n(t) = \sum_{j=-\infty}^{n-1} \sum_{k=-\infty}^{\infty} \widehat{d}_{j,k}\psi_{j,k}(t) \qquad (5.3.3)$$

we have, again,

$$\widehat{d}_{j,k} = (W_{\widetilde{\psi}}f_n)\left(\frac{k}{2^j}, \frac{1}{2^j}\right), \qquad (5.3.4)$$

but now since $f_n(t) \in V_n$, the roles of $\psi(t)$ and $\widetilde{\psi}(t)$ *cannot* be interchanged unless $\widetilde{V}_n = V_n$ or, equivalently, $\psi(t)$ is a semiorthogonal wavelet (see (5.1.35)–(5.1.36)).

Let $\phi(t)$ be any compactly supported scaling function as described by (5.2.1), and let $\psi(t)$ be its corresponding semiorthogonal wavelet with minimum support as given by (5.2.22). Then, by Theorem 5.3 the frequency-domain description of the two-scale relations of $\phi(t)$ and $\psi(t)$ is given by

$$\begin{cases} \widehat{\phi}(\omega) = P(z)\widehat{\phi}\left(\dfrac{\omega}{2}\right) \quad \text{and} \\[2mm] \widehat{\psi}(\omega) = Q(z)\widehat{\phi}\left(\dfrac{\omega}{2}\right), \qquad z = e^{-j\omega/2}, \end{cases} \qquad (5.3.5)$$

with

$$Q(z) = -z^{2N+1} P\left(-\frac{1}{z}\right) E_\phi(-z),$$ (5.3.6)

where N is defined by (5.2.21), and we have chosen the constant c in (5.2.20) to be -1.

Let us now study the duals $\widetilde{\phi}(t)$ and $\widetilde{\psi}(t)$ of $\phi(t)$ and $\psi(t)$, respectively, in some detail. For $\widetilde{\phi}(t)$, we recall from (5.1.20) and (5.1.21) that its Fourier transform is given by

$$\widehat{\widetilde{\phi}}(\omega) = \frac{1}{E_\phi(z^2)} \widehat{\phi}(\omega), \qquad z = e^{-j\omega/2}.$$ (5.3.7)

So, as already described by (5.1.27), $\widetilde{\phi}(t)$ generates the same MRA $\{V_j\}$ as $\phi(t)$. As for $\widetilde{\psi}(t)$, being semiorthogonal, it generates the same orthogonal complementary subspaces $\{W_j\}$ as $\psi(t)$. Hence, it makes sense also to study the two-scale relations of $\widetilde{\phi}(t)$ and $\widetilde{\psi}(t)$. This is one of the objectives of this section. Let $\{a_k\}$ and $\{b_k\}$ denote their corresponding two-scale sequences with two-scale symbols

$$\begin{cases} A(z) := \dfrac{1}{2} \displaystyle\sum_k a_k z^k, \\[2mm] B(z) := \dfrac{1}{2} \displaystyle\sum_k b_k z^k. \end{cases}$$ (5.3.8)

That is, in analogy to (5.3.5), the frequency-domain formulation of the two-scale relations of $\widetilde{\phi}(t)$ and $\widetilde{\psi}(t)$ is given by

$$\begin{cases} \widehat{\widetilde{\phi}}(\omega) = A(z)\widehat{\widetilde{\phi}}\left(\dfrac{\omega}{2}\right) \quad \text{and} \\[2mm] \widehat{\widetilde{\psi}}(\omega) = B(z)\widehat{\widetilde{\phi}}\left(\dfrac{\omega}{2}\right), \qquad z = e^{-j\omega/2}. \end{cases}$$ (5.3.9)

To determine $A(z)$ and $B(z)$, we need the following identities.

Theorem 5.4. *Let $P(z)$ and $Q(z)$ be the two-scale symbols of a compactly supported scaling function $\phi(t)$ and its corresponding minimally supported wavelet $\psi(t)$, and let $E_\phi(z)$ be the E–F L-polynomial associated with $\phi(t)$. Then*

$$E_\phi(z^2) = |P(z)|^2 E_\phi(z) + |P(-z)|^2 E_\phi(-z),$$ (5.3.10)

and, for $z = e^{-j\omega/2}$,

$$\sum_k |\widehat{\psi}(\omega + 2\pi k)|^2 = |Q(z)|^2 E_\phi(z) + |Q(-z)|^2 E_\phi(-z)$$ (5.3.11)

$$= E_\phi(z) E_\phi(-z) E_\phi(z^2).$$

The proof of (5.3.10) is straightforward, simply by first breaking up the summation in (5.1.21) into two parts, one being the sum over all even k and the other being the sum over all odd k and then by applying the first identity in (5.3.5). The same trick, with the only exception that the second instead of the first identity in (5.3.5) is used, also gives the first identity in (5.3.11). To establish the second identity in (5.3.11), we simply apply (5.3.6) and (5.3.10) consecutively.

As an application of (5.3.11) in Theorem 5.4, we see that the duality relation (5.1.39) can be reformulated as

$$\widehat{\widetilde{\psi}}(\omega) = \frac{1}{E_\phi(z)E_\phi(-z)E_\phi(z^2)}\widehat{\psi}(\omega), \tag{5.3.12}$$

and by (5.3.5) and (5.3.6), this formula reduces to

$$\widehat{\widetilde{\psi}}(\omega) = \frac{-z^{2N+1}P\left(-\frac{1}{z}\right)}{E_\phi(z)E_\phi(z^2)}\widehat{\phi}\left(\frac{\omega}{2}\right). \tag{5.3.13}$$

Now, by a straightforward application of (5.3.7), (5.3.13), and the two-scale relations (5.3.5) and (5.3.9), we can even compute the two-scale sequence $\{a_k\}$ and $\{b_k\}$ of the dual scaling function $\widetilde{\phi}(t)$ and dual wavelet $\widetilde{\psi}(t)$ as follows. A more general formulation is given in section 6.2 of Chapter 6.

Theorem 5.5. *The two-scale symbols $A(z)$ and $B(z)$ in (5.3.8)–(5.3.9) are given by*

$$A(z) = \frac{E_\phi(z)}{E_\phi(z^2)}P(z) \tag{5.3.14}$$

and

$$B(z) = \frac{-z^{2N+1}}{E_\phi(z^2)}P\left(-\frac{1}{z}\right), \tag{5.3.15}$$

where N is defined as in (5.2.21).

Furthermore, there is a duality relationship between the two pairs of two-scale symbols: $(P(z), Q(z))$ and $(A(z), B(z))$. This identity can be easily translated into the corresponding duality relationship between the two-scale sequences $(\{p_k\}, \{q_k\})$ of $\phi(t)$, $\psi(t)$ and the two-scale sequences $(\{a_k\}, \{b_k\})$ of their duals $\widetilde{\phi}(t)$, $\widetilde{\psi}(t)$. This result, which can be easily established by using the identity (5.3.10) in Theorem 5.4, is stated as follows.

Theorem 5.6. *The two-scale relations of $\phi(t)$ and its semiorthogonal wavelet $\psi(t)$ and the two-scale relations of their corresponding duals $\widetilde{\phi}(t)$ and $\widetilde{\psi}(t)$ are governed by the duality identity*

$$P(z)\overline{A(z)} + Q(z)\overline{B(z)} = 1, \qquad |z| = 1, \tag{5.3.16}$$

or, equivalently,

$$\sum_{k=0}^{M}p_k a_{k-n} + \sum_{k=0}^{M+2n_\phi}q_k b_{k-n} = 4\delta_{n,0}, \qquad n \in \mathbb{Z}. \tag{5.3.17}$$

In (5.3.17), since p_k and q_k were assumed to be real valued, we have dropped the complex conjugation sign over the real constants a_k and b_k.

It can be easily verified that the two-scale symbols satisfy the identity

$$P(z)\overline{A(-z)} + Q(z)\overline{B(-z)} = 0, \qquad |z| = 1 \qquad (5.3.18)$$

as well. The important common characteristic of the identities (5.3.16) and (5.3.18) is that they are both "symmetric," in the sense that the pairs $(P(z), Q(z))$ and $(\overline{A(z)}, \overline{B(z)})$ can be interchanged.

While the first pair describes the two-scale relations of $\phi(t)$ and $\psi(t)$, the second pair governs their decomposition relation as follows.

Theorem 5.7. *The decomposition relation of $\phi(t)$ and its semiorthogonal wavelet $\psi(t)$ is given by*

$$\phi(2t - \ell) = \sum_k \{a_{\ell-2k}\phi(t - k) + b_{\ell-2k}\psi(t - k)\}, \qquad \ell \in \mathbb{Z}. \qquad (5.3.19)$$

Here, again, if complex-valued p_k and q_k were considered, then we need complex conjugation of the *decomposition sequences* $\{a_k\}$ and $\{b_k\}$ in (5.3.19). The reason they are called decomposition sequences will be clear in a moment.

To establish (5.3.19), we consider its frequency-domain formulation

$$\hat{\phi}\left(\frac{\omega}{2}\right) z^n = \left(\sum_k a_{n-2k} z^{2k}\right) \hat{\phi}(\omega) + \left(\sum_k b_{n-2k} z^{2k}\right) \hat{\psi}(\omega), \qquad z = e^{-j\omega/2},$$

or, equivalently,

$$\hat{\phi}\left(\frac{\omega}{2}\right) = (\overline{A(z)} \pm \overline{A(-z)})\hat{\phi}(\omega) + (\overline{B(z)} \pm \overline{B(-z)})\hat{\psi}(\omega), \qquad z = e^{-j\omega/2}, \qquad (5.3.20)$$

where the notation of the two $\pm$ signs means that (5.3.20) is valid if either both positive or both negative signs are used. The proof of (5.3.20) follows immediately from (5.3.16) and (5.3.18) after the two-scale relations in (5.3.5) are applied.

Let us now explain why the sequences $\{a_k\}$ and $\{b_k\}$ in (5.3.19) are called *decomposition sequences.*

Theorem 5.8. *Let*

$$f_n(t) = \sum_k c_{n,k}\phi(2^n t - k) \qquad (5.3.21)$$

in V_n be written as

$$f_n(t) = f_{n-1}(t) + g_{n-1}(t), \qquad (5.3.22)$$

where $f_{n-1}(t) \in V_{n-1}$ and $g_{n-1}(t) \in W_{n-1}$. Then the coefficients $\{c_{n-1,k}: k \in \mathbb{Z}\}$ and $\{d_{n-1,k}: k \in \mathbb{Z}\}$ of the decomposed components

$$\begin{cases} f_{n-1}(t) = \sum_k c_{n-1,k}\phi(2^{n-1}t - k) \quad and \\ \\ g_{n-1}(t) = \sum_k d_{n-1,k}\psi(2^{n-1}t - k) \end{cases} \qquad (5.3.23)$$

can be computed by applying the "decomposition algorithm":

$$\begin{cases} c_{n-1,k} = \sum_{\ell} a_{\ell-2k} c_{n,\ell}, \\ d_{n-1,k} = \sum_{\ell} b_{\ell-2k} c_{n,\ell}. \end{cases} \quad (5.3.24)$$

On the other hand, the two-scale sequences $\{p_k\}$ and $\{q_k\}$ are also called *reconstruction sequences*, because of the following.

Theorem 5.9. *Let $f_n(t) \in V_n$ be given by*

$$f_n(t) = f_{n-1}(t) + g_{n-1}(t), \quad (5.3.25)$$

where $f_{n-1}(t) \in V_{n-1}$ and $g_{n-1}(t) \in W_{n-1}$, and write

$$\begin{cases} f_{n-1}(t) = \sum_{k} c_{n-1,k} \phi(2^{n-1}t - k), \\ g_{n-1}(t) = \sum_{k} d_{n-1,k} \psi(2^{n-1}t - k). \end{cases} \quad (5.3.26)$$

Then the coefficient sequence $\{c_{n,k} : k \in \mathbb{Z}\}$ of the signal

$$f_n(t) = \sum_{k} c_{n,k} \phi(2^n t - k),$$

reconstructed from its components $f_{n-1}(t)$ and $g_{n-1}(t)$ as in (5.3.25), can be computed by applying the following "reconstruction algorithm":

$$c_{n,k} = \sum_{\ell} \{p_{k-2\ell} c_{n-1,\ell} + q_{k-2\ell} d_{n-1,\ell}\}. \quad (5.3.27)$$

Observe that because both filter sequences $\{p_k\}$ and $\{q_k\}$ in (5.3.27) are finite, the moving-average (MA) process (5.3.27) is simply the sum of two finite impulse response (FIR) filters. Of course, the "input sequences" $\{c_{n-1,k} : k \in \mathbb{Z}\}$ and $\{d_{n-1,k} : k \in \mathbb{Z}\}$ must first be "upsampled" before the FIR filters are applied. However, although the decomposition sequences $\{a_k\}$ and $\{b_k\}$ in (5.3.24) have exponential decay, they are infinite sequences and, therefore, require either truncation or ARMA infinite impulse response (IIR) implementation. Fortunately, in view of the duality relationship as stated in Theorem 5.6, we can use the finite sequences $\{p_{-k}\}$ and $\{q_{-k}\}$ for wavelet decomposition instead. This is the following so-called "duality principle."

Theorem 5.10. *Let $f_n(t) \in V_n$ be given by*

$$f_n(t) = \sum_{k} \tilde{c}_{n,k} \tilde{\phi}(2^n t - k) = \sum_{k} \tilde{c}_{n-1,k} \tilde{\phi}(2^{n-1}t - k) + \sum_{k} \tilde{d}_{n-1,k} \tilde{\psi}(2^{n-1}t - k).$$

$$(5.3.28)$$

Then the sequences $\{\widetilde{c}_{n-1,k}\colon\ k\in\mathbb{Z}\}$ and $\{\widetilde{d}_{n-1,k}\colon\ k\in\mathbb{Z}\}$ can be computed by applying the "decomposition algorithm"

$$\begin{cases} \widetilde{c}_{j-1,k} = \displaystyle\sum_{\ell} p_{2k-\ell}\,\widetilde{c}_{j,\ell}, \\[2mm] \widetilde{d}_{j-1,k} = \displaystyle\sum_{\ell} q_{2k-\ell}\,\widetilde{c}_{j,\ell}. \end{cases} \tag{5.3.29}$$

We recall that if the scaling function $\phi(t)$ is orthonormal, so that the semiorthogonal wavelet $\psi(t)$ becomes an orthonormal wavelet, then the reconstruction sequence $\{q_k\}$ and decomposition sequences $\{a_k\}$, $\{b_k\}$ reduce to

$$\begin{cases} q_k = (-1)^k p_{1-k}, \\[2mm] a_k = \dfrac{1}{2}\,p_{-k}, \\[2mm] b_k = \dfrac{1}{2}\,q_{-k} = \dfrac{1}{2}\,(-1)^k p_{k+1}, \end{cases} \tag{5.3.30}$$

as given by (4.4.32)–(4.4.33). In this orthonormal setting, a rescaling by $\frac{1}{\sqrt{2}}$ as introduced in (4.4.34) somewhat facilitates computer implementation. However, when semiorthogonal wavelets are considered, such a rescaling factor is not recommended, particularly for the cardinal B-wavelets $\psi_m(t)$ whose reconstruction sequences $\{p_k\}$ and $\{q_k\}$, as given in (5.2.7) and (5.2.24), respectively, are very simple rational numbers.

Example 5.5. Let $N_m(t)$ be the mth-order B-spline, where $m \geq 1$ is fixed. Then its dual $\widetilde{N}_m(t)$, which was already introduced in Definition 3.5, is determined by (5.3.7); that is,

$$\widehat{N}_m(\omega) = \left(\sum_{k=-m+1}^{m-1} N_{2m}(m+k)e^{-jk\omega}\right)\widehat{\widetilde{N}}_m(\omega) \tag{5.3.31}$$

or, equivalently,

$$N_m(t) = \sum_{n=-m+1}^{m-1} N_{2m}(m+n)\widetilde{N}_m(t-n). \tag{5.3.32}$$

Hence, any B-spline series representation

$$f_n(t) = \sum_{k} c_{n,k} N_m(2^n t - k) \tag{5.3.33}$$

as in (5.3.21) can be written as a dual series

$$f_n(t) = \sum_{k} \widetilde{c}_{n,k} \widetilde{N}_m(2^n t - k) \tag{5.3.34}$$

as in (5.3.28), with

$$\widetilde{c}_{n,k} = \sum_{p=1}^{2m-1} N_{2m}(p)c_{n,m+k-p}. \qquad (5.3.35)$$

Observe that the change to dual representation only requires a very simple FIR filter. With this change of bases, we can then apply the duality principle stated in Theorem 5.10 and use the finite sequences $\{p_{-k}\}$ and $\{q_{-k}\}$, where

$$\begin{cases} p_k = 2^{-m+1}\dbinom{m}{k}, & k = 0, \ldots, m, \\[2mm] q_k = \dfrac{(-1)^k}{2^{m-1}} \displaystyle\sum_{\ell=0}^{m} \dbinom{m}{\ell} N_{2m}(k-\ell+1), & k = 0, \ldots, 3m-2, \end{cases} \qquad (5.3.36)$$

as filter sequences in the decomposition algorithm (5.3.29).

To change back to a B-spline series representation, however, we need to invert the formula in (5.3.32). To do so, we set

$$G(z) = \sum_n g_n z^n := \frac{1}{E_\phi(z)} = \frac{1}{\sum_{n=-m+1}^{m-1} N_{2m}(m+n)z^n} \qquad (5.3.37)$$

(with $\phi(t) = N_m(t)$) and obtain

$$\widetilde{N}_m(t) = \sum_n g_n N_m(t-n). \qquad (5.3.38)$$

Similarly, it also follows from (5.3.12) that the change of wavelet bases can be accomplished by

$$\widetilde{\psi}_m(t) = \sum_n \left\{ \sum_{p,q} (-1)^p g_p g_q g_{2n-2q-p} \right\} \psi_m(t-n) \qquad (5.3.39)$$

and

$$\psi_m(t) = \sum_n \left\{ \sum_{p,q} (-1)^p N_{2m}(m+p)N_{2m}(m+q)N_{2m}(m+2n-2q-p) \right\}$$
$$\times \widetilde{\psi}_m(t-n). \qquad (5.3.40)$$

Finally, the decomposition sequences $\{a_k\}$ and $\{b_k\}$ as defined in (5.3.8)–(5.3.9) can be easily written down by applying (5.3.14)–(5.3.15) and (5.3.37) as follows.

$$\begin{cases} a_n = 2^{-m+1} \displaystyle\sum_{\ell=0}^{m} \dbinom{m}{\ell} \sum_{2k=n-\ell-m+1}^{n-\ell+m-1} N_{2m}(m+n-2k-\ell)g_k, \\[4mm] b_n = (-1)^{n+1} 2^{-m+1} \displaystyle\sum_{2k=n-2m}^{n-m} \dbinom{m}{n-2k-m} g_k. \end{cases} \qquad (5.3.41)$$

The graphs of the dual B-wavelets $\tilde{\psi}_m(t)$ and their filter characteristics for $m = 2, \ldots, 6$ are shown in Figures 5.7–5.11.

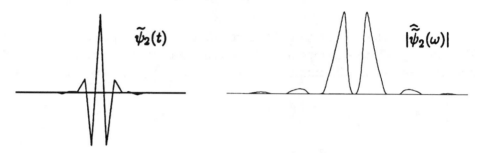

FIG. 5.7. *Linear dual B-wavelet $\tilde{\psi}_2(t)$ and its filter characteristic.*

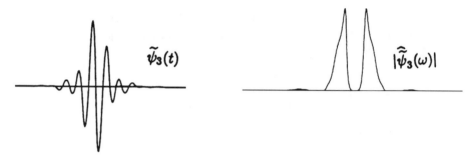

FIG. 5.8. *Quadratic dual B-wavelet $\tilde{\psi}_3(t)$ and its filter characteristic.*

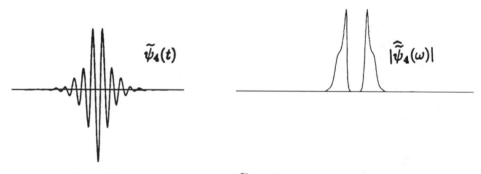

FIG. 5.9. *Cubic dual B-wavelet $\tilde{\psi}_4(t)$ and its filter characteristic.*

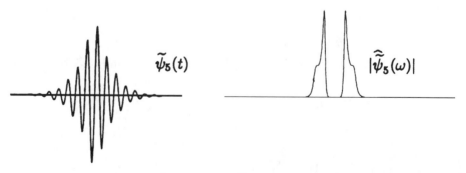

FIG. 5.10. *Quartic dual B-wavelet $\tilde{\psi}_5(t)$ and its filter characteristic.*

FIG. 5.11. *Quintic dual B-wavelet $\widetilde{\psi}_6(t)$ and its filter characteristic.*

In Table 5.5, we list the values of $a_n := a_{m;n}$ and $b_n := b_{m;n}$ for linear $(m = 2)$ and cubic $(m = 4)$ B-wavelet decompositions.

TABLE 5.5. *Decomposition sequences for linear $(m = 2)$ and cubic $(m = 4)$ B-wavelet decompositions. Observe that $a_k = a_{m-k}$ and $b_k = b_{3m-2-k}$.*

	$m = 2$		$m = 4$	
k	a_k	b_{k+1}	a_{k+1}	b_{k+4}
1	0.683012701892	0.866025403784	0.893162856314	-1.475394519892
2	0.316987298108	-0.316987298108	0.400680825467	0.468422596633
3	-0.116025403784	-0.232050807569	-0.282211870811	0.742097698478
4	-0.084936490539	0.084936490539	-0.232924626134	-0.345770890775
5	0.031088913246	0.062177826491	0.129083571218	-0.389745580800
6	0.022758664048	-0.022758664047	0.126457446356	0.196794277304
7	-0.008330249198	-0.016660498395	-0.066420837387	0.207690838380
8	-0.006098165652	0.006098165652	-0.067903608499	-0.106775803373
9	0.002232083545	0.004464167091	0.035226101674	-0.111058440711
10	0.001633998562	-0.001633998561	0.036373586989	0.057330952254
11	-0.000598084983	-0.001196169967	-0.018815686621	0.059433388390
12	-0.000437828595	0.000437828595	-0.019473269356	-0.030709700871
13	0.000160256388	0.000320512777	0.010066747520	-0.031811811318
14	0.000117315818	-0.000117315818	0.010424052187	0.016440944687
15	-0.000042940569	-0.000085881139	-0.005387929819	0.017028029466
16	-0.000031434679	0.000031434678	-0.005579839208	-0.008800839839
17	0.000011505891	0.000023011782	0.002883979478	-0.009114745138
18	0.000008422897	-0.000008422897	0.002986784625	0.004710957034
19	-0.000003082990	-0.000006165980	-0.001543728719	0.004878941541
20	-0.000002256905	0.0000022569054	-0.001598768083	-0.002521687975
21	0.000000826079	0.0000016521587	0.000826326663	-0.002611601542

5.4. Total positivity and optimality of time-frequency windows.

As we mentioned in Chapter 2 (see (2.2.7) and (2.4.20)), for both low-pass and bandpass time-frequency localization, the time-frequency windows are governed by the uncertainty lower bound of $\frac{1}{2}$. While the value $\frac{1}{2}$ is sharp for both lowpass and bandpass filters in the sense that it is the greatest lower bound, it cannot be achieved by any scaling function (for lowpass filtering) and any wavelet (for bandpass filtering). We have also seen in the previous section (see Theorems 5.1 and 5.2) that higher-order orthonormal scaling functions and wavelets have large time-frequency windows. Recall also that for lowpass filtering, the Gaussian function (see (2.2.8)) attains the uncertainty lower bound $\frac{1}{2}$. Hence, it is natural to believe that scaling functions $\phi(t)$ with the Gaussian shape should give near optimal (i.e., $\Delta_\phi \Delta_{\widehat{\phi}}$ is close to $\frac{1}{2}$) time-frequency localization, while wavelets $\psi(t)$ with graphs close to the modulated Gaussian (i.e., multiplication of a Gaussian by the sine or cosine functions) have a fair chance of having near optimal bandpass time-frequency localization performance (i.e., $\Delta_\psi \Delta_{\widehat{\psi}}^+$ is close to $\frac{1}{2}$). That this is indeed the case is explained by a very important algebraic concept called *total positivity* (TP). It will be seen that not only do the Gaussian functions have this nice TP property, but many other functions, including the cardinal B-splines, are totally positive (TP) as well. Consequently, we will see that the mth-order cardinal B-splines $N_m(t)$ satisfy

$$\Delta_{N_m} \Delta_{\widehat{N}_m} \to \frac{1}{2} \quad \text{as} \quad m \to \infty, \tag{5.4.1}$$

and the corresponding mth-order cardinal B-wavelets $\psi_m(t)$, as defined in (5.2.24)–(5.2.25), also satisfy

$$\Delta_{\psi_m} \Delta_{\widehat{\psi}_m}^+ \to \frac{1}{2} \quad \text{as} \quad m \to \infty. \tag{5.4.2}$$

Values of $\Delta_{\psi_m} \Delta_{\widehat{\psi}_m}^+$ for $m = 2, \ldots, 11$ were given in Table 5.4. Observe that even for the cubic case, we already have

$$\frac{\Delta_{\psi_4} \Delta_{\widehat{\psi}_4}^+ - 0.5}{0.5} = .0095372,$$

which is less than 1% away from being optimal. For completeness, we include a table of the $\Delta_{N_m} \Delta_{\widehat{N}_m}$ values for the cardinal B-splines in Table 5.6.

TABLE 5.6.   *Values of* $\Delta_{N_m} \Delta_{\widehat{N}_m}$ *of the B-splines.*

m	$\Delta_{N_m} \Delta_{\widehat{N}_m}$	m	$\Delta_{N_m} \Delta_{\widehat{N}_m}$
2	0.547675918	7	0.500342152
3	0.50382161	8	0.500257489
4	0.501230696	9	0.500203424
5	0.50070501	10	0.500163058
6	0.500474053	11	0.500132981

Let us first introduce the notion of TP matrices.

Definition 5.2. *A matrix A (finite, infinite, or bi-infinite) is TP if all its minors (i.e., determinants of finite square submatrices formed by deleting arbitrary rows and columns of A) are nonnegative.*

Observe that since any entry of a matrix is one of its minors, a TP matrix must have nonnegative entries. Of course most matrices with nonnegative entries are not TP. We will be interested in those TP matrices that are induced by a sequence or by a function, as in the following two definitions.

Definition 5.3. *A sequence $\{p_k\}$ (finite, infinite, or bi-infinite) is called a Pólya frequency (PF) sequence if the bi-infinite Toeplitz matrix $A = [a_{i,j}]$, defined by $a_{i,j} = p_{i-j}$ (where we set $p_\ell := 0$ if the index ℓ is not in the domain of definition of the given sequence $\{p_k\}$), is totally positive.*

Typical examples of PF sequences include the binomial coefficients as follows.

Example 5.6. Let m be any positive integer and $p_k = \binom{m}{k} = \frac{m!}{(m-k)!k!}$ for $k = 0, \ldots, m$. Then $\{p_k\}$ is a PF sequence.

In this example, we set $p_\ell = 0$ for $\ell < 0$ or $\ell > m$. The corresponding TP Toeplitz matrix is given by $A = [a_{i,j}]$ with $a_{i,j} = p_{i-j}$.

Definition 5.4. *A function $F(t, y)$ of two variables is called a TP kernel, if the matrix $A = [F(t_i, y_j)]$ is a TP matrix, where $\{t_i\}$ and $\{y_j\}$ are arbitrarily chosen increasing sequences (with (t_i, y_j) in the domain of definition of $F(t, y)$). In addition, a function $f(t)$, continuous for all real t, is called a TP function if the function $F(t, k)$, defined by*

$$F(t, k) := f(t - k), \qquad k \in \mathbb{Z},$$

is a TP kernel.

Example 5.7. The mth-order cardinal B-splines $N_m(t)$, $m \geq 1$, are TP functions.

Observe that there is some connection between the above two examples. Indeed, the two-scale sequence of $N_m(t)$ is $\{2^{-m+1}\binom{m}{k}\}$, which is a PF sequence by Example 5.6. That Example 5.7 actually follows from Example 5.6 is a consequence of the following results.

Theorem 5.11. *Let $\phi(t)$ be a scaling function with a two-scale sequence $\{p_k\}$ as in Definition 3.3. Then $\phi(t)$ is a TP function, provided that $\{p_k\}$ is a finite PF sequence.*

If, in addition, the PF two-scale sequences are symmetric, then we will see that their scaling functions have near-optimal time-frequency localization, provided that their (approximation) orders are sufficiently high. More precisely, we have the following.

Theorem 5.12. *Let $1 \le n_1 < n_2 < \cdots$ be positive integers that satisfy*

$$n_m = O(m) \tag{5.4.3}$$

(meaning that $\frac{n_m}{m}$ is bounded as $m \to \infty$). Suppose that $\Phi_m(t)$ are scaling functions with two-scale sequences $\{p_{m,k}\}_k$ that satisfy the following four conditions:

(i) *$p_{m,0} \ne 0$, but $p_{m,\ell} = 0$ for $\ell < 0$ or $\ell > n_m$;*
(ii) *$\{p_{m,k}\}_k$ are symmetric, i.e.,*

$$p_{m,k} = p_{m,n_m-k}, \quad k = 0, \ldots, n_m; \tag{5.4.4}$$

(iii) *$\{p_{m,k}\}_k$ are PF sequences; and*
(iv) *the zero $z = -1$ of the two-scale symbol*

$$P_m(z) = \frac{1}{2} \sum_{k=0}^{n_m} p_{m,k} z^k \tag{5.4.5}$$

is of order m, meaning that

$$P_m(z) = \left(\frac{1+z}{2}\right)^m S_m(z), \tag{5.4.6}$$

where $S_m(-1) \ne 0$ and $S_m(1) = 1$.

Then the scaling functions $\Phi_m(t)$ have the asymptotically optimal time-frequency localization property, meaning that

$$\Delta_{\Phi_m} \Delta_{\widehat{\Phi}_m} \to \frac{1}{2} \quad as \quad m \to \infty. \tag{5.4.7}$$

Of course, if $n_m = m$ then $\Phi_m(t) = N_m(t)$ is the mth-order cardinal B-spline. The localization property of $N_m(t)$ has already been demonstrated in Table 5.6.

If the two-scale sequences $\{p_{m,k}\}_k$ of some scaling functions $\Phi_m(t)$ satisfy (i)–(iv), we will call $\Phi_m(t)$ *stoplets* (where s stands for symmetry and t o p for total positivity), and we will also call m (the order of the zero $z = -1$ of the symbol $P_m(z)$ in (5.4.5)) the order of the stoplet $\Phi_m(t)$. Hence, the above theorem says that mth-order stoplets have the asymptotically optimal time-frequency localization property.

In the previous section, we have outlined a general procedure for constructing all semiorthogonal wavelets $\psi(t)$ that correspond to a given scaling function $\phi(t)$. In particular, by choosing the integer shift N in (5.2.21) and constant $c = -1$ for the polynomial $Q(z)$ in (5.2.20), this wavelet $\psi(t)$ is uniquely determined by $\phi(t)$. In the following, we will use the notation

$$\Psi_m(t) \tag{5.4.8}$$

to denote the semiorthogonal wavelet with minimum support, uniquely determined by a given mth-order stoplet $\Phi_m(t)$ with shift N and constant c as
described above. Since $\Psi_m(t)$ has minimum time duration while having the
same number of vanishing moments as the other wavelets in the same class as
described by (5.2.14), we call $\Psi_m(t)$ the mth-order *cowlet* (where c o stands for
complete oscillation, and wlet for wavelet) that corresponds to the mth-order
stoplet $\Phi_m(t)$.

Example 5.8. By choosing $n_m = m$ in Theorem 5.12, the mth-order stoplet
$\Phi_m(t)$ and cowlet $\Psi_m(t)$ become the mth-order cardinal B-spline $N_m(t)$ and
B-wavelet $\psi_m(t)$, respectively.

For $n_m > m$, since the polynomial factor $S_m(z)$ in (5.4.6) is of degree
$n_m - m$, there is some freedom for improving the "filters" $\Phi_m(t)$ and $\Psi_m(t)$
for lowpass and bandpass time-frequency localization, respectively. In the following, we see that cowlets, in general, also have the asymptotically optimal
time-frequency localization property.

Theorem 5.13. *Let $\Psi_m(t)$ be the cowlets in (5.4.8) that correspond to some
mth-order stoplets $\Phi_m(t)$. Then under conditions (i)–(iv) in Theorem 5.12,
$\Psi_m(t)$ have the localization property*

$$\Delta_{\Psi_m} \Delta_{\widehat{\Psi}_m}^+ \;\to\; \frac{1}{2} \quad as \quad m \to \infty. \tag{5.4.9}$$

We end this chapter by mentioning that since stoplets $\Phi_m(t)$ are symmetric
and totally positive, they are very desirable scaling functions for analog signal
representation. For instance, the number of sign changes of the continuous-
time signal

$$f(t) = \sum_k c_k \Phi_m(2^n t - k) \tag{5.4.10}$$

(where the integer n is fixed) does not exceed the number of sign changes of
the coefficient sequence $\{c_k\}$. Hence, if $c_k \geq 0$ for all k, then $f(t) \geq 0$ for all
t. Also, when the MRA $\{V_n\}$ generated by $\Phi_m(t)$ locally contain all quadratic
polynomials, then $f(t)$ is nondecreasing and/or convex if the sequence $\{c_k\}$ is
nondecreasing and/or convex. In addition, since $\Phi_m(t)$ has finite time duration,
these geometric properties of $f(t)$ are "controlled" by the coefficient sequence
$\{c_k\}$ locally. The points $(\frac{k+n_m/2}{2^n}, c_k)$ are called "control points" of the graph
of the function $f(t)$ in (5.4.10).

CHAPTER **6**

Algorithms

One of the main reasons the wavelet transform is considered to be a very powerful mathematical tool for signal analysis is its fast algorithms both for computation and for implementation. This chapter is devoted to a discussion of the most basic algorithms, including those for signal representations and display. In addition, several variations of the decomposition and reconstruction algorithms will be studied, and boundary scaling functions and wavelets for eliminating boundary effects will be introduced. This consideration includes B-splines and B-wavelets with arbitrary knots on a bounded interval for modeling and analyzing discrete data information taken at nonuniform sample points.

6.1. Signal representations.

The integral wavelet transform (IWT) is defined for analyzing analog signals with finite energy. Hence, if only discrete samples of such a signal $f(t)$ are available, we have to map these digital samples to the analog domain in order for the IWT to be meaningful. In fact, even if continuous-time information of the signal $f(t)$ is known, it still has to be mapped into some multiresolution analysis (MRA) space V_n before any of the fast algorithms can be applied. In other words, we are interested in mapping a signal $f(t)$ to one of the form

$$f_n(t) = \sum_k c_{n,k} \phi(2^n t - k) \tag{6.1.1}$$

for some scaling function $\phi(t)$, which generates an MRA $\{V_j\}$ of L^2. The importance of the signal representation (6.1.1), as we recall from Theorem 5.8 in Chapter 5 (see also (1.3.8)–(1.3.11) and Figure 1.21), is that the decomposition algorithm in (1.3.11) or (5.3.24) can be applied not only to separate $f_n(t)$ into a DC (or blur) component $f_{n-1}(t)$ and an AC (or wavelet) component $g_{n-1}(t)$ but also to determine the wavelet transform of the representation $f_n(t)$ at the time-scale positions $\left(\frac{k}{2^{n-1}}, \frac{1}{2^{n-1}}\right)$, $k \in \mathbb{Z}$ using the dual wavelet $\tilde{\psi}(t)$ as the analyzing wavelet. The reader is reminded of (5.1.17) with $j = n - 1$, where $\{d_{n-1,k}\}$ is the coefficient sequence of the wavelet series expansion of the AC component $g_{n-1}(t)$. This section is concerned with mapping the signal $f(t)$ to $f_n(t) \in V_n$ using perhaps partial information about the signal.

If $f(t)$ is known for all t (or on a sufficiently dense discrete set of the time domain), then orthogonal projection (or L^2-projection) of the signal $f(t)$ to the space V_n is the most natural choice. This is done by taking the inner product of $f(t)$ with respect to the "dual" $2^n \widetilde{\phi}(2^n t - k)$, where $\widetilde{\phi}(t) \in V_0$ is the dual scaling function of $\phi(t)$ as defined in (5.1.19). That is, the coefficients in the series expansion (6.1.1) are given by

$$c_{n,k} = 2^n \int_{-\infty}^{\infty} f(t)\widetilde{\phi}(2^n t - k)dt. \qquad (6.1.2)$$

Observe that if $\widetilde{\phi}(t)$ has compact support (i.e., finite time duration), the integral in (6.1.2) is a definite integral. Of course in actual computation of the integral, a numerical integration (quadrature) scheme is used, and this induces errors. When the "resolution" is not very high (i.e., n is not large enough), the error of numerical computation of $c_{n,k}$ in (6.1.2) by using the data information $f\left(\frac{k}{2^n}\right)$, $k \in \mathbb{Z}$, may be unacceptably large. In this and many other situations, it is usually better to rely on other methods to find the coefficients $c_{n,k}$, so that the series representation $f_n(t)$ in (6.1.1) is a desirable analog model of the discrete data $f\left(\frac{k}{2^n}\right)$, $k \in \mathbb{Z}$. One of the most effective methods is "interpolation," which means that the coefficients $c_{n,k}$ are chosen so that the analog representation $f_n(t)$ agrees with the discrete data $f\left(\frac{k}{2^n}\right)$ at $t = \frac{k}{2^n}$; namely,

$$f_n\left(\frac{k}{2^n}\right) = f\left(\frac{k}{2^n}\right), \qquad k \in \mathbb{Z}. \qquad (6.1.3)$$

As a simple example, recall from Figure 3.6 that the linear B-spline $N_2(t)$ has the interpolating property

$$N_2(k) = \delta_{k,1}, \qquad k \in \mathbb{Z}; \qquad (6.1.4)$$

i.e., the centered B-spline

$$L_2(t) := N_2(t+1) \qquad (6.1.5)$$

is a fundamental cardinal interpolating spline (see (4.3.7) in Chapter 4, with $m = 1$). In general, as we have already seen in section 4.3 of Chapter 4, however, higher-order fundamental cardinal interpolating splines $L_{2m}(t)$, $m \geq 2$ have infinite support and can be formulated as

$$\begin{cases} L_{2m}(t) = \sum_k c_k^{(2m)} N_{2m}(t + m - k), \text{ with} \\[2mm] \sum_k c_k^{(2m)} z^k = \dfrac{1}{E_m(z)} = \dfrac{1}{\sum_k N_{2m}(m+k)z^k}, \end{cases} \qquad (6.1.6)$$

where $E_m(z)$ is the Euler–Frobenius Laurent polynomial (E–F L-polynomial) associated with the $2m$th-order cardinal B-spline $N_{2m}(t)$. Observe that $E_1(z) \equiv 1$

and (6.1.6) reduces to (6.1.5) when $2m = 2$ or $m = 1$. With $L_{2m}(t)$, the spline interpolating model of any digital signal can be readily obtained as follows.

Example 6.1. Let $f\left(\frac{k}{2^n}\right)$, $k \in \mathbb{Z}$, be any given discrete data set, where n is a (fixed) integer. Then the $(2m)$th-order cardinal spline

$$f_n(t) = \sum_k c_{n,k} N_{2m}(2^n t - k) \tag{6.1.7}$$

that satisfies the interpolating property (6.1.3) is simply

$$f_n(t) = \sum_k f\left(\frac{k}{2^n}\right) L_{2m}(2^n t - k). \tag{6.1.8}$$

Hence, by applying (6.1.6), it follows that the coefficient sequence $\{c_{n,k}\}$ $k \in \mathbb{Z}$, is given by

$$c_{n,k} = \sum_\ell c_\ell^{(2m)} f\left(\frac{(k+m) - \ell}{2^n}\right). \tag{6.1.9}$$

In particular, for linear splines (i.e., $2m = 2$) since $c_\ell^{(2)} = \delta_{\ell,0}$, we have $c_{n,k} = f\left(\frac{k+1}{2^n}\right)$ in (6.1.7). This is reassured, of course, by (6.1.5) and (6.1.8) with $2m = 2$.

In mapping a digital signal $f\left(\frac{k}{2^n}\right)$, $k \in \mathbb{Z}$ to an analog representation $f_n(t)$ with scaling function $\phi(t)$, there is always the temptation to use the given data values $f\left(\frac{k}{2^n}\right)$ as coefficients $c_{n,k}$; namely,

$$f_n(t) = \sum_k f\left(\frac{k}{2^n}\right) \phi(2^n t - k). \tag{6.1.10}$$

If $\widetilde{\phi}(t)$ is a high-order Butterworth lowpass filter as defined by (2.3.4), then we have seen in (3.1.29) of Chapter 3 that indeed for signals $f(t)$ with sufficiently *small* bandwidth, the representation (6.1.10) is acceptable, provided that $\widetilde{\phi}(t)$ is centered at the origin. In general, such a representation is not recommended since the wavelet coefficients $d_{j,k}$, $j < n$ and $k \in \mathbb{Z}$, which are $2^{j/2}$ multiples of the IWT of $f_n(t)$, are not very effective in revealing certain features such as oscillation of the digital signal $f\left(\frac{k}{2^n}\right)$, $k \in \mathbb{Z}$.

Example 6.2. Let $n = 0$ and consider the data $f(k) = \delta_{k,-1} - \delta_{k,0} + \delta_{k,1}$, $k \in \mathbb{Z}$ as shown in Figure 6.1a. By using these data values as coefficients in the analog model with the centered cubic B-spline $N_4(t+2)$ as the scaling function, we have

$$f_0(t) = \sum_k f(k) N_4(t+2-k)$$

$$= N_4(t+3) - N_4(t+2) + N_4(t+1),$$

and this function takes on values

$$f_0(k) = \frac{1}{6}\delta_{k,-2} + \frac{1}{2}\delta_{k,-1} - \frac{1}{3}\delta_{k,0} + \frac{1}{2}\delta_{k,1} + \frac{1}{6}\delta_{k,2}$$

at the integers. These digital samples are shown in Figure 6.1b, and the analog representation $f_0(t)$ itself is shown in Figure 6.1c. Observe that the digital samples of the model $f_0(t)$ have much lower frequency than that of the original digital data at the origin. In other words the high frequency content of the original signal is not well represented by $f_0(t)$.

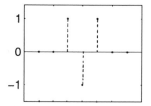 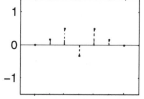 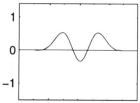

FIG. 6.1a. *Original signal.*   FIG. 6.1b. *Digital samples*   FIG. 6.1c. *Analog model.*
of analog model.

Unfortunately, for cubic spline interpolation the weight sequence $\{c_k^{(4)}\}$ in (6.1.6) and (6.1.9) (with $2m = 4$) is bi-infinite, although the decay at $\pm\infty$ is exponentially fast. Furthermore, although the error induced by truncation is not substantial, the effect is that cubic polynomials are no longer reproduced by the (truncated) interpolant. As a consequence, those DWT values $d_{j,k}$, which are supposed to vanish for local cubic polynomial data, are not equal to zero anymore. A study of their closeness to zero in terms of the number of terms of the truncated weight sequence $\{c_k^{(4)}\}$, $|k| \leq N$, is worthwhile. For this and other purposes, we give the values of $c_k^{(4)}$ in the following.

Example 6.3. Let $\{c_k^{(4)}\}$ be the weight sequence in (6.1.6) with $m = 2$ for computing cubic spline interpolations $f_n(t)$ in (6.1.8) (by the discrete convolution procedure (6.1.9)). Then by (6.1.6), its z-transform (or symbol) is given by

$$\sum_k c_k^{(4)} z^k = \frac{1}{E_2(z)} = \frac{1}{\sum_k N_4(2+k)z^k}$$

$$= \frac{6}{z^{-1} + 4 + z} = \frac{6}{(z^{-1} + z_0)(1 + z_0^{-1}z)}, \qquad (6.1.11)$$

where $z_0 = 2 - \sqrt{3}$. This gives, for $|z| = 1$,

$$\sum_k c_k^{(4)} z^k = \sqrt{3} \left\{ \frac{1}{1 + z_0 z} + \frac{-1}{1 + z_0^{-1}z} \right\}$$

$$= \sqrt{3} \left\{ \frac{1}{1 + z_0 z} + \frac{-z_0 z^{-1}}{1 + z_0 z^{-1}} \right\}$$

$$= \sqrt{3} \left\{ \sum_{k=0}^{\infty} (-1)^k z_0^k z^k + (-z_0 z^{-1}) \sum_{k=0}^{\infty} (-z_0 z^{-1})^k \right\}$$

$$= \sqrt{3} \sum_{k=-\infty}^{\infty} (-1)^k z_0^{|k|} z^k. \tag{6.1.12}$$

Hence, we have

$$c_k^{(4)} = (-1)^k \sqrt{3} \left(2 - \sqrt{3} \right)^{|k|}, \qquad k \in \mathbb{Z}. \tag{6.1.13}$$

Observe that the z-transform of $\{c_k^{(4)}\}$ is a rational function, which gives rise to recursive filtering. For stability, however, since $z_0 = 2 - \sqrt{3}$ lies inside $|z| < 1$, the factor $(1 + z_0^{-1} z)$ in (6.1.11) must be changed to $1 + z_0 z^{-1}$. Hence, we may either apply the "transfer function"

$$\sum_k c_k^{(4)} z^k = 6 z_0 \frac{1}{1 + z_0 z^{-1}} \cdot \frac{1}{1 + z_0 z},$$

as given by (6.1.11), in cascade: first with unit advances z and then with unit delays z^{-1} or in parallel, as given by (6.1.12); namely,

$$\sum_k c_k^{(4)} z^k = \frac{\sqrt{3}}{1 + z_0 z} + \frac{-\sqrt{3} z_0 z^{-1}}{1 + z_0 z^{-1}},$$

one from left to right and the other from right to left. Of course these two methods of implementation require knowledge of the full data set as well as initial values.

The above cascade implementation can also be explained by means of matrix factorization as follows.

Example 6.4. The discrete convolution operation (6.1.9) with $2m = 4$ can be expressed as multiplication of the bi-infinite column data vector

$$\left[\cdots \ f\left(\frac{1}{2^n}\right) \ f\left(\frac{2}{2^n}\right) \ f\left(\frac{3}{2^n}\right) \ \cdots \right]$$

(with $f\left(\frac{2}{2^n}\right)$ as center) by the inverse of the bi-infinite Toeplitz matrix

$$E := \begin{bmatrix} \ddots & & \ddots & & & & \\ \ddots & & \ddots & \ddots & & \bigcirc & \\ & & \frac{1}{6} & \frac{4}{6} & \frac{1}{6} & & \\ & \bigcirc & & \ddots & \ddots & \ddots & \\ & & & & \ddots & \ddots & \ddots \end{bmatrix} \tag{6.1.14}$$

(with $\frac{4}{6}$ in the main diagonal). Now, the matrix E has LU decomposition

$$E = \frac{2+\sqrt{3}}{6} LL^T \tag{6.1.15}$$

where L^T is the transpose of the lower triangular Toeplitz matrix

$$L = \begin{bmatrix} & \ddots & & & \\ \ddots & & \ddots & & \bigcirc \\ & 2-\sqrt{3} & & 1 & \\ & & \ddots & & \ddots \\ \bigcirc & & & \ddots & \ddots \end{bmatrix}$$

with 1 in the main diagonal and $2 - \sqrt{3}$ in the off diagonal, and these two diagonals are the only nonzero ones. Hence, inversion of the matrix E (for computing the B-spline coefficient sequence $\{c_{n,k}\}$, $k \in \mathbb{Z}$ in (6.1.9)) only requires a forward difference operation followed by a backward difference operation.

Another method for avoiding truncation of the weight sequences $\{c_k^{(2m)}\}$ is to "approximately" interpolate the data. This operation, called *quasi interpolation*, must preserve all the polynomials contained locally in the MRA, so that the discrete wavelet transform (DWT), $\widehat{d}_{j,k}$ of local polynomial data sets (of order $2m$) vanish. For example, in the cubic spline setting we require the quasi-interpolation operator to preserve all cubic polynomials. While cubic spline quasi interpolation is not unique, we give the one with the shortest filter in the following.

Example 6.5. The operator

$$(Q_n f)(t) = \frac{1}{6} \sum_k \left\{ -f\left(\frac{k+1}{2^n}\right) + 8f\left(\frac{k+2}{2^n}\right) - f\left(\frac{k+3}{2^n}\right) \right\} N_4(2^n t - k),$$
$$\tag{6.1.16}$$

which maps the data set $\{f(\frac{k}{2^n})\}$ to the MRA subspace V_n (generated by the cardinal cubic B-spline), preserves all cubic polynomials locally. Hence, if the efficient map $(Q_n f)(t)$ in (6.1.16) is used, the DWT, $\widehat{d}_{j,k}$, of this analog model vanish for all local cubic polynomial data sets.

6.2. Orthogonal decompositions and reconstructions.

Although the basic algorithms for wavelet decompositions and reconstruction have been introduced in both Chapter 1 (see (1.3.11) and (1.3.12) along with Figures 1.21–1.23) and Chapter 5 (see Theorems 5.8 and 5.9), they were designed only for bi-infinite data sets and multilevel decompositions that are governed by scale $a = \frac{1}{2}$. In order to take care of other situations such as bounded intervals, multidimensional data sets, and spline models with arbitrary knots, etc., in this section we give a more general framework of the decomposition and reconstruction scheme. However, since our main concern is to gain flexibility in interchanging the analysis and synthesis (wavelet) filters, we restrict our attention to orthogonal decompositions (see section 5.1). This consideration, of course, includes semiorthogonal and, particularly, orthonormal wavelets (see (5.1.36)–(5.1.37)).

Let U be any (finite- or infinite-dimensional) Hilbert space with inner product $\langle , \rangle$ and basis $\{u_i\}$, and let V be a proper subspace of U. Assume that we are given a basis $\{v_i\}$ of V along with the matrix P that governs the relation of this basis with the given basis $\{u_i\}$ of U; namely,

$$\begin{cases} v_i = \sum_j p_{i,j} u_j, \\ P = [p_{i,j}]. \end{cases} \tag{6.2.1}$$

In the following, we give a characterization of all bases of the orthogonal complementary subspace W of U relative to V; i.e.,

$$\begin{cases} U = V \oplus W, \text{ which means} \\ U = V + W \text{ and } V \perp W. \end{cases} \tag{6.2.2}$$

For this purpose, we need the Gramian matrix

$$G = [\langle u_i, u_j \rangle] \tag{6.2.3}$$

of the given basis of U. It is well known that the matrix G is Hermitian and invertible.

Theorem 6.1. *Let $\{w_i\} \subset U$ and write*

$$\begin{cases} w_i = \sum_j q_{i,j} u_j, \\ Q = [q_{i,j}]. \end{cases} \tag{6.2.4}$$

Then $\{w_i\}$ is a basis of W if and only if

$$\begin{cases} Q G \overline{P}^T = 0 \text{ and} \\ Q G \overline{Q}^T > 0. \end{cases} \tag{6.2.5}$$

Example 6.6. If $\phi(t)$ is a scaling function with a two-scale sequence $\{p_k\}$, then we have

$$\phi(t - i) = \sum_j p_{j-2i}\phi(2t - j). \tag{6.2.6}$$

Hence, by considering $U = V_1$ and $V = V_0$, where $\{V_n\}$ is the MRA of L^2 generated by $\phi(t)$, we have $P = [p_{i,j}]$ with

$$p_{i,j} = p_{j-2i}. \tag{6.2.7}$$

If, in addition, $\phi(t)$ is an orthonormal scaling function, then the Gramian matrix G is a $\frac{1}{2}$ multiple of the identity matrix (with the dimension given by that of U). Recall that by setting $q_k = (-1)^k \bar{p}_{1-k}$ (see (4.1.18) and the remark on complex-valued p_k's that follows the formula), the family

$$\psi(t - i) = \sum_j (-1)^j \bar{p}_{2i-j+1} \phi(2t - j) \qquad (6.2.8)$$

constitutes an orthonormal basis of the orthogonal complementary subspace $W = W_0$ of U relative to V. That is, we have $Q = [q_{i,j}]$ with

$$q_{i,j} = (-1)^j \bar{p}_{2i-j+1}. \qquad (6.2.9)$$

Observe that the (i, j)th entry of $PG\overline{Q}^T$ is given by

$$\sum_\ell p_{\ell-2i}(-1)^\ell p_{2j-\ell+1}$$

$$= \sum_k (-1)^k p_k p_{2(i+j)-k+1},$$

which has been shown to be zero in section 4.1 (see the verification of (4.1.18)). That is, we have

$$QG\overline{P}^T = \overline{PG\overline{Q}^T} = 0.$$

To justify the second statement in (6.2.5), we observe that the (i, j)th entry of $QG\overline{Q}^T = Q\overline{Q}^T$ is given by

$$\sum_k q_{i,k} \bar{q}_{j,k} = \sum_k (-1)^k \bar{p}_{2i-k+1}(-1)^k p_{2j-k+1}$$

$$= \sum_\ell \bar{p}_{2i-\ell} p_{2j-\ell},$$

and that the orthonormality condition (4.4.9) on the two-scale symbol $P(z)$ is equivalent to

$$\sum_\ell \bar{p}_{2i-\ell} p_{2j-\ell} = \delta_{i,j}, \qquad i, j \in \mathbb{Z}. \qquad (6.2.10)$$

Example 6.7. As a continuation of the above example, we recall the notion of autocorrelation functions $\Phi(t)$ in (5.2.2) of scaling functions $\phi(t)$ in Chapter 5. It follows that the Gramian matrix G in (6.2.3) is given by

$$G = \frac{1}{2}[\Phi(i - j)]. \qquad (6.2.11)$$

In particular, for the mth-order cardinal B-spline $\phi(t) = N_m(t)$, we have

$$G = \frac{1}{2}[N_{2m}(m + i - j)], \qquad (6.2.12)$$

and as a consequence of (3.4.7) and (6.2.7) we also have $P = [p_{i,j}]$ with

$$p_{i,j} = 2^{-m+1} \binom{m}{j - 2i}, \tag{6.2.13}$$

where $\binom{m}{\ell} := 0$ for $\ell < 0$ or $\ell > m$. We remark that the Gramian matrix (6.2.12) replaces the E–F L-polynomial $E_m(z) = E_{N_m}(z)$ in (5.2.6). Again, for the mth-order cardinal spline setting, it follows from (5.2.24) that the matrix $Q = [q_{i,j}]$ in (6.2.4) is given by

$$q_{i,j} = \frac{(-1)^j}{2^{m-1}} \sum_{\ell=0}^{m} \binom{m}{\ell} N_{2m}(j + 1 - 2i - \ell). \tag{6.2.14}$$

Returning to the general Hilbert space setting (6.2.1)–(6.2.5), we assume, for the rest of this section, that (6.2.5) is satisfied. Then the Gramian matrices

$$\begin{cases} H = [\langle v_i, v_j \rangle] \quad \text{and} \\ K = [\langle w_i, w_j \rangle] \end{cases} \tag{6.2.15}$$

are also Hermitian and invertible. The decomposition relation (6.2.2), described by

$$u_i = \sum_j a_{i,j} v_j + \sum_j b_{i,j} w_j, \tag{6.2.16}$$

is uniquely determined by the three Gramian matrices G, H, K and the "two-scale" matrices P, Q as follows.

Theorem 6.2. *Let*

$$A = [a_{ij}] \quad and \quad B = [b_{ij}] \tag{6.2.17}$$

be the decomposition matrices as introduced in (6.2.16). Then

$$\begin{cases} A = G\overline{P}^T H^{-1}, \\ B = G\overline{Q}^T K^{-1}. \end{cases} \tag{6.2.18}$$

With the matrices A and B in (6.2.18), we can decompose any element

$$F = \sum_k e_k u_k \tag{6.2.19}$$

in U as an orthogonal sum of

$$\begin{cases} f = \sum_k c_k v_k \quad \text{and} \\ g = \sum_k d_k w_k \end{cases} \tag{6.2.20}$$

in V and W, respectively, simply by matrix multiplications, as in the following.

Algorithm 6.1 (decomposition algorithm). *Let*

$$
\begin{cases}
\mathbf{e} = [\dots e_k \dots]^T, \\
\mathbf{c} = [\dots c_k \dots]^T, \quad \text{and} \\
\mathbf{d} = [\dots d_k \dots]^T
\end{cases}
\tag{6.2.21}
$$

be the column vectors that represent the elements $F, f,$ and g of $U, V,$ and W, respectively, as given by (6.2.19)–(6.2.20). Then

$$
\begin{cases}
\mathbf{c} = A^T \mathbf{e}, \\
\mathbf{d} = B^T \mathbf{e}.
\end{cases}
\tag{6.2.22}
$$

On the other hand, the following algorithm is an immediate consequence of the two-scale relations (6.2.1) and (6.2.4).

Algorithm 6.2 (reconstruction algorithm). *The sum $F = f + g$ as in (6.2.19)–(6.2.20) can be computed by using*

$$
\mathbf{e} = P^T \mathbf{c} + Q^T \mathbf{d},
\tag{6.2.23}
$$

where $\mathbf{e}, \mathbf{c}, \mathbf{d}$ are the column vectors in (6.2.21).

Example 6.8. As a further continuation of Example 6.6, if the scaling function $\phi(t)$ is orthonormal, then the Gramian matrices H and K are identity matrices, and G is a $\frac{1}{2}$ multiple of the identity. Hence, it follows from Theorem 6.2 that

$$
A = \frac{1}{2}\overline{P}^T \quad \text{and} \quad B = \frac{1}{2}\overline{Q}^T
\tag{6.2.24}
$$

so that, by (6.2.7) and (6.2.9), we have

$$
\begin{cases}
a_{i,j} = \dfrac{1}{2}\overline{p}_{i-2j}, \\[2mm]
b_{i,j} = \dfrac{(-1)^i}{2}\overline{p}_{2j-i+1}.
\end{cases}
\tag{6.2.25}
$$

By using (6.2.25), it can be easily verified that the orthogonal decomposition (4.1.19), with orthonormal scaling functions $\phi(t)$ and orthonormal wavelets $\psi(t)$ in Chapter 4, is a special case of (6.2.16) if we set $u_i = \phi(2t - i)$, $v_j = \phi(t - j)$, and $w_j = \psi(t - j)$. Observe that the downsampling and upsampling operations are implicit in the matrix multiplications (6.2.22) and (6.2.23), respectively.

Returning to the general Hilbert space setting (6.2.1)–(6.2.5), we next demonstrate that the decomposition matrices A, B can be interchanged with the reconstruction matrices P, Q. This is clear, at least intuitively, from the following "duality principle." For the sake of clarity, we need the subscripts in

$$
I_U, \quad I_V, \quad I_W
\tag{6.2.26}
$$

to identify the identity operators (matrices) of the Hilbert spaces U, V, W, respectively, onto themselves.

Theorem 6.3 (duality principle). *The matrices $A, B, P,$ and Q satisfy the relations*

$$PA = I_V, \quad QB = I_W, \tag{6.2.27}$$

and

$$AP + BQ = I_U. \tag{6.2.28}$$

To explain what (6.2.27) means, we consider the basis $\{\tilde{u}_i\}$ of U which is dual to $\{u_i\}$ in the sense that

$$\langle u_i, \tilde{u}_j \rangle = \delta_{i,j}. \tag{6.2.29}$$

We will call $\{\tilde{u}_i\}$ the *dual basis* relative to $\{u_i\}$ for U. Then by introducing

$$\begin{cases} \tilde{v}_i = \sum_j \bar{a}_{j,i} \tilde{u}_j, \\ \tilde{w}_i = \sum_j \bar{b}_{j,i}, \tilde{u}_j, \end{cases} \tag{6.2.30}$$

we see that $\{\tilde{v}_i\}$ is the dual basis relative to $\{v_i\}$ for V, and $\{\tilde{w}_i\}$ is the dual basis relative to $\{w_i\}$ for W. Hence, if these dual bases are used as bases for $U, V,$ and W, then the matrices

$$\bar{A}^T \quad \text{and} \quad \bar{B}^T$$

are the reconstruction matrices in Algorithm 6.2 (in place of P and Q, respectively). On the other hand, the relation (6.2.28) also implies that the matrices

$$\bar{P}^T \quad \text{and} \quad \bar{Q}^T$$

are the decomposition matrices in Algorithm 6.1 (in place of A and B, respectively), and that the decomposition relation (6.2.16) is replaced by

$$\tilde{u}_i = \sum_j \bar{p}_{j,i} \tilde{v}_j + \sum_j \bar{q}_{j,i} \tilde{w}_j. \tag{6.2.31}$$

It is easy to see that the Gramian matrices $G, H,$ and K in (6.2.3) and (6.2.15) can be used to relate the given bases to their respective duals as follows:

$$\begin{cases} u_i = \sum_j \langle u_i, u_j \rangle \tilde{u}_j, \\ v_i = \sum_j \langle v_i, v_j \rangle \tilde{v}_j, \\ w_i = \sum_j \langle w_i, w_j \rangle \tilde{w}_j. \end{cases} \tag{6.2.32}$$

Let us now return to examine the essence of the structure of the matrices A and B in (6.2.18) for the purpose of giving other variations of the decomposition algorithm in (6.2.22). By using (6.2.32) and (6.2.18), we can break up the decomposition algorithm (6.2.22) into three steps; namely,

$$\begin{cases} \mathbf{c} = \overline{H}^{-1}\widetilde{\mathbf{c}}, \\ \mathbf{d} = \overline{K}^{-1}\widetilde{\mathbf{d}}, \end{cases} \tag{6.2.33a}$$

where

$$\begin{cases} \widetilde{\mathbf{c}} = \overline{P}\widetilde{\mathbf{e}}, \\ \widetilde{\mathbf{d}} = \overline{Q}\widetilde{\mathbf{e}}, \end{cases} \tag{6.2.33b}$$

and

$$\widetilde{\mathbf{e}} = \overline{G}\mathbf{e}. \tag{6.2.33c}$$

By using the notations $\widetilde{\mathbf{e}} = [\ldots \widetilde{e}_k \ldots]^T$, $\widetilde{\mathbf{c}} = [\ldots \widetilde{c}_k \ldots]^T$, and $\widetilde{\mathbf{d}} = [\ldots \widetilde{d}_k \ldots]^T$, we see that (6.2.33a) and (6.2.33c) are algorithms for the change of dual bases in the representations

$$\begin{cases} F = \sum_k e_k u_k = \sum_k \widetilde{e}_k \widetilde{u}_k, \\ f = \sum_k c_k v_k = \sum_k \widetilde{c}_k \widetilde{v}_k, \\ g = \sum_k d_k w_k = \sum_k \widetilde{d}_k \widetilde{w}_k, \end{cases} \tag{6.2.34}$$

and that (6.2.33b) says that $\overline{P}$ and $\overline{Q}$ (instead of A^T and B^T is (6.2.22)) are used as decomposition matrices.

Example 6.9. Let $m \geq 1$ and consider the MRA $\{V_n\}$ generated by the mth-order cardinal B-spline $N_m(t)$. Then considering $U = V_1$, $V = V_0$, and $W = W_0$, it follows from (5.3.32) and (5.3.40) that the Gramian matrices H and K in (6.2.33a)–(6.2.33b) are given by

$$\begin{cases} H = \overline{H} = [N_{2m}(m+i-j)], \\ K = \overline{K} = \left[\sum_{p,q}(-1)^p N_{2m}(m+p)N_{2m}(m+q)N_{2m}(m+2(i-j)-2q-p)\right] \end{cases} \tag{6.2.35}$$

with i and j denoting the row and column indices, respectively. Of course, the Gramian matrix G in (6.2.33c) is given by

$$G = \overline{G} = \frac{1}{2}H. \tag{6.2.36}$$

The matrices P and Q in (6.2.33b) have been discussed in Example 6.7 (see (6.2.13)–(6.2.14)).

This change of bases, from $N_m(t)$ to $\widetilde{N}_m(t)$, reflects the use of $N_m(t)$ (instead of $\widetilde{N}_m(t)$) and $\psi_m(t)$ (instead of $\widetilde{\psi}_m(t)$) for both signal representation and filtering. The significance is not only that $\Delta_{N_m}\Delta_{\widehat{N}_m}$ and $\Delta_{\psi_m}\Delta_{\widehat{\psi}_m}^{+}$ are asymptotically optimal, as described by Theorems 5.12–5.13, but also that the filter characteristics of $\widehat{N}_m(\omega)$ and $\widehat{\psi}_m(\omega)$, as shown in Figures 3.6–3.8 and 5.2–5.6, are more Gaussian-like than those of their duals, as shown in Figures 3.9–3.11 and 5.7–5.11. Furthermore, the finite sequences $\{p_k\}$ and $\{q_k\}$ given in (3.4.7) and (5.2.24) are used for both decomposition and reconstruction.

6.3. Graphical display of signal representations.

Let us return to the setting of an MRA $\{V_n\}$ of L^2 with scaling function $\phi(t)$ and wavelet $\psi(t)$ as described in (1.2.1)–(1.2.2) and (1.2.6)–(1.2.7). In section 6.1, we discussed several procedures for mapping a signal $f(t)$ to an MRA space V_n in the form of

$$f_n(t) = \sum_k c_{n,k}\phi(2^n t - k), \qquad (6.3.1)$$

so that the decomposition and reconstruction algorithms (1.3.11)–(1.3.12) can be applied to the coefficient sequence representation $\{c_{n,k}\}$, $k \in \mathbb{Z}$, of the signal $f_n(t)$. After the sequence $\{c_{n,k}\}$ (or a modification of it as a result of such applications as data compression) has been reconstructed, it is (usually) necessary to compute $f_n(t)$ in (6.3.1) by using the values of $c_{n,k}$. In some applications, it is also desirable to display the graph of $f_n(t)$ directly from the values of $c_{n,k}$. Several efficient algorithms for the cardinal B-splines $N_m(t)$ and B-spline series representations are available in the literature. However, we are not going into any of them but only mention that the recursive algorithm

$$\begin{cases} N_m(t) = \dfrac{t}{m-1}N_{m-1}(t) + \dfrac{m-t}{m-1}N_{m-1}(t-1), \\ \text{with } N_1(t) = 1 \text{ for } 0 \le t < 1, \text{ and } 0 \text{ otherwise,} \end{cases} \qquad (6.3.2)$$

as given in Theorem 3.3, can be used not only for computing $N_m(t)$ for each t but also for finding the explicit polynomial pieces of $N_m(t)$ on the interval $[k, k+1]$, $k = 0, \ldots, m-1$. (Recall that $N_m(t) = 0$ for $t < 0$ or $t > m$.)

For the Daubechies scaling functions $\phi_{D;m}(t)$, however, the only information we have is their two-scale sequences. In the following, we give three methods for computing $\phi_{D;m}(t)$ from this information. To simplify the notation, let us consider

$$\phi(t) = \sum_{k=0}^{n} p_k\phi(2t - k), \text{ with } p_0 p_n \ne 0. \qquad (6.3.3)$$

Method 1. Let

$$P(z) = \frac{1}{2}\sum_{k=0}^{n} p_k z^k$$

be the two-scale symbol of $\phi(t)$. Then $\phi(t)$ can be computed by taking the inverse Fourier transform of the infinite product

$$\prod_{k=1}^{\infty} P(e^{-i\omega/2^k}). \tag{6.3.4}$$

The fast Fourier transform (FFT) can be used for this purpose.

Method 2. Compute $\phi(t)$ on a "dense set" $2^{-N}\mathbb{Z}$ (i.e., for a sufficiently large positive integer N). This can be done recursively as follows. First compute $\phi(t)$ for $t = k \in \mathbb{Z}$ by solving the equation

$$\phi(k) = \sum_{\ell=0}^{n} p_\ell \phi(2k - \ell) = \sum_{\ell} p_{2k-\ell} \phi(\ell) \tag{6.3.5}$$

for $\{\phi(k)\}$ that satisfies the normalization condition

$$\sum_{k} \phi(k) = \sum_{k=0}^{n} \phi(k) = 1. \tag{6.3.6}$$

Here, we have used the support property

$$\text{supp } \phi = [0, n], \tag{6.3.7}$$

which is a consequence of (6.3.3). For all practical purposes (such as $\phi(t) = \phi_{D;m}(t)$ for all $m \geq 2$), we have continuous $\phi(t)$, so that $\phi(0) = \phi(n) = 0$. From the values $\phi(k)$, $k \in \mathbb{Z}$, we next compute

$$\phi\left(\frac{k}{2}\right) = \sum_{\ell=0}^{n} p_\ell \phi(k - \ell), \qquad k \in \mathbb{Z},$$

and from these values we compute

$$\phi\left(\frac{k}{2^2}\right) = \sum_{\ell=0}^{n} p_\ell \phi\left(\frac{k}{2} - \ell\right),$$

etc., and, finally,

$$\phi\left(\frac{k}{2^N}\right) = \sum_{\ell=0}^{n} p_\ell \phi\left(\frac{k}{2^{N-1}} - \ell\right).$$

Since the two-scale sequence $\{p_\ell\}$ satisfies

$$\sum_{\ell=0}^{n} p_\ell = 2, \tag{6.3.8}$$

the normalization condition (6.3.6) guarantees that

$$\frac{1}{2^N} \sum_k \phi\left(\frac{k}{2^N}\right) = \frac{1}{2^N} \sum_{k=0}^{2^N n} \phi\left(\frac{k}{2^N}\right) = 1. \tag{6.3.9}$$

Observe that the quantity in (6.3.9) is a Riemann sum approximation of the integral

$$\int_{-\infty}^{\infty} \phi(t)dt = \int_0^n \phi(t)dt, \tag{6.3.10}$$

which is supposed to be 1 for any scaling function.

Method 3. Set

$$\phi_0(t) = N_2(t),$$

where $N_2(t)$ is the linear cardinal B-spline, and compute

$$\phi_k(t) = \sum_{\ell=0}^n p_\ell \phi_{k-1}(2t - 1) \tag{6.3.11}$$

for $k = 1, 2, \dots$. Then

$$\phi(t) = \lim_{k \to \infty} \phi_k(t).$$

Observe that since

$$\int_{-\infty}^{\infty} N_2(t)dt = \int_0^2 N_2(t)dt = 1,$$

it follows from (6.3.8) that

$$\int_{-\infty}^{\infty} \phi_k(t)dt = 1$$

for all $k = 1, 2, \dots$, so that the limit function $\phi(t)$ satisfies (6.3.10) also. We remark that although the choice of $N_2(t)$ as the initial function $\phi_0(t)$ is not essential, any initial function $\phi_0(t)$, however, must satisfy

$$\int_{-\infty}^{\infty} \phi_0(t)dt = 1$$

for the iterative scheme (6.3.11) to converge to $\phi(t)$.

6.4. Multidimensional wavelet transforms.

The IWT (or CWT) defined for functions of one variable can be easily extended to the s-dimensional setting, where $s \geq 2$, by considering

$$(W_{\psi_1,\dots,\psi_s} f)(b_1, \dots, b_s; a_1, \dots, a_s)$$
$$:= \frac{1}{\sqrt{a_1 \dots a_s}} \int_{-\infty}^{\infty} \cdots \int_{-\infty}^{\infty} f(t_1, \dots, t_s)$$
$$\times \psi_1\left(\frac{t_1 - b_1}{a_1}\right) \dots \psi_s\left(\frac{t_s - b_s}{a_s}\right) dt_1 \dots dt_s. \tag{6.4.1}$$

For simplicity, we only study the two-dimensional setting with $\psi_1(t) = \psi_2(t) = \widetilde{\psi}(t)$. That is, we consider

$$(W_{\widetilde{\psi}}f)(b_1, b_2; a_1, a_2) = \frac{1}{\sqrt{a_1 a_2}} \int_{-\infty}^{\infty} \int_{-\infty}^{\infty} f(x,y) \overline{\widetilde{\psi}\left(\frac{x-b_1}{a_1}\right) \widetilde{\psi}\left(\frac{y-b_2}{a_2}\right)} \, dx \, dy,$$

$$(6.4.2)$$

where $f(x,y)$ has finite (two-dimensional) energy and $\widetilde{\psi}(t)$ is an analyzing wavelet. As in the one-dimensional case, the DWT that corresponds to the IWT in (6.4.2) is given by

$$\widehat{d}_{j_1, j_2; k_1, k_2} := (W_{\widetilde{\psi}}f)\left(\frac{k_1}{2^{j_1}}, \frac{k_2}{2^{j_2}}; \frac{1}{2^{j_1}}, \frac{1}{2^{j_2}}\right), \qquad j_1, j_2, k_1, k_2 \in \mathbb{Z}. \qquad (6.4.3)$$

Since each of the variables x and y of $f(x,y)$ is mapped to its own frequency (octave) domain in terms of scales 2^{-j_1} and 2^{-j_2}, respectively, we call $\widehat{d}_{j_1, j_2; k_1, k_2}$ the multifrequency (or, more precisely, bifrequency) DWT of $f(x,y)$.

As before, let $\phi(t)$ be a scaling function that generates some MRA of $L^2 = L^2(-\infty, \infty)$, and $\psi(t)$ be a corresponding wavelet that generates certain direct-sum complementary subspaces of this MRA with dual wavelet $\widetilde{\psi}(t)$, so that (5.1.1)–(5.1.9) are satisfied. Then by using the dual wavelet $\widetilde{\psi}(t)$ as the analyzing wavelet in the multifrequency DWT in (6.4.3), we see that every (two-dimensional) finite-energy signal $f(x,y)$ has a (unique) wavelet series expansion

$$f(x,y) = \sum_{j_1, j_2, k_1, k_2 \in \mathbb{Z}} \widehat{d}_{j_1, j_2; k_1, k_2} \Psi_{j_1, j_2; k_1, k_2}(x,y), \qquad (6.4.4)$$

where

$$\Psi_{j_1, j_2; k_1, k_2}(x,y) := 2^{(j_1+j_2)/2} \Psi(2^{j_1} x - k_1, 2^{j_2} y - k_2), \qquad (6.4.5)$$

with $\Psi(x,y)$ given by the "tensor product"

$$\Psi(x,y) := \psi(x)\psi(y) \qquad (6.4.6)$$

of $\psi(x)$ and $\psi(y)$. We remark in passing that although nonseparable (i.e., nontensor product) multidimensional wavelets $\Psi(x,y)$ are available, we only restrict our attention to the tensor-product ones as defined in (6.4.6). One immediate consequence of this restriction is that by using the usual notations for $\psi_{j,k}(t)$ and $\widetilde{\psi}_{j,k}(t)$ as in (5.1.4) and (5.1.6), the series representation (6.4.4) can be written as

$$f(x,y) = \sum_{j_1, j_2, k_1, k_2} \widehat{d}_{j_1, j_2; k_1, k_2} \psi_{j_1, k_1}(x) \psi_{j_2, k_2}(y). \qquad (6.4.7)$$

In applications, we only consider RMS bandlimited signals (see section 2.2). Hence, we approximate the finite-energy signal $f(x,y)$ in (6.4.7) by partial sums of the form

$$f_{n_1, n_2}(x,y) := \sum_{j_1=-\infty}^{n_1-1} \sum_{j_2=-\infty}^{n_2-1} \sum_{k_1, k_2 \in \mathbb{Z}} \widehat{d}_{j_1, j_2; k_1, k_2} \psi_{j_1, k_1}(x) \psi_{j_2, k_2}(y). \qquad (6.4.8)$$

Let us first consider the special case where $n_1 = n_2 = n$ and set

$$f_n(x, y) := f_{n,n}(x, y). \tag{6.4.9}$$

Then by (6.4.8) we have

$$f_n(x, y) = f_{n-1}(x, y) + g_{n-1}(x, y) \tag{6.4.10}$$

where

$$g_{n-1}(x, y) := g_{LH,n-1}(x, y) + g_{HL,n-1}(x, y) + g_{HH,n-1}(x, y) \tag{6.4.11}$$

with

$$\begin{cases} g_{HH,n-1}(x, y) = \displaystyle\sum_{k_1,k_2 \in \mathbb{Z}} d^{(n-1)}_{HH,k_1,k_2} \psi(2^{n-1}x - k_1)\psi(2^{n-1}y - k_2), \\[2mm] d^{(n-1)}_{HH,k_1,k_2} := 2^{(n-1)/2}\widehat{d}_{n-1,n-1;k_1,k_2}; \end{cases} \tag{6.4.12}$$

$$g_{LH,n-1}(x, y) := \sum_{k_2 \in \mathbb{Z}} \left\{ \sum_{j_1=-\infty}^{n-2} \sum_{k_1 \in \mathbb{Z}} \widehat{d}_{j_1,n-1;k_1,k_2} \phi_{j_1,k_1}(x) \right\} \psi_{n-1,k_2}(y)$$

$$=: \sum_{k_1,k_2 \in \mathbb{Z}} d^{(n-1)}_{LH,k_1,k_2} \phi(2^{n-1}x - k_1)\psi(2^{n-1}y - k_2); \tag{6.4.13}$$

and

$$g_{HL,n-1}(x, y) := \sum_{k_1 \in \mathbb{Z}} \left\{ \sum_{j_2=-\infty}^{n-2} \sum_{k_2 \in \mathbb{Z}} \widehat{d}_{n-1,j_2;k_1,k_2} \phi_{j_2,k_2}(y) \right\} \psi_{n-1,k_1}(x)$$

$$=: \sum_{k_1,k_2 \in \mathbb{Z}} d^{(n-1)}_{HL,k_1,k_2} \psi(2^{n-1}x - k_1)\phi(2^{n-1}y - k_2). \tag{6.4.14}$$

We remark that in (6.4.13) and (6.4.14) we have used the fact that

$$V_{n-1} = W_{n-2} + W_{n-3} + \cdots \tag{6.4.15}$$

(applied to the variable x in (6.4.13) and the variable y in (6.4.14)). The indices L and H are used to indicate low- and high-frequency components. Hence, HH stands for high frequencies in both x and y directions and LH for low frequencies in the x direction and high frequencies in the y direction. For convenience, we may also apply (6.4.15) twice (first to the variable x and then to the variable y) to $f_n(x, y)$ and $f_{n-1}(x, y)$ in (6.4.8)–(6.4.10) and write

$$f_j(x, y) = \sum_{k_1,k_2 \in \mathbb{Z}} c^{(j)}_{LL,k_1,k_2} \phi(2^j x - k_1)\phi(2^j x - k_2) \tag{6.4.16}$$

for $j = n, n - 1$. Then, as in the one-dimensional setting, the wavelet decomposition algorithm is applied to the coefficient sequence $\{c_{LL,k_1,k_2}^{(n)}\}$ to yield $\{c_{LL,k_1,k_2}^{(n-1)}\}$, $\{d_{LH,k_1,k_2}^{(n-1)}\}$, $\{d_{HL,k_1,k_2}^{(n-1)}\}$, and $\{d_{HH,k_1,k_2}^{(n-1)}\}$ that represent $f_{n-1}(x,y)$, $g_{LH,n-1}(x,y)$, $g_{HL,n-1}(x,y)$, and $g_{HH,n-1}(x,y)$ in (6.4.16), (6.4.13), (6.4.14), and (6.4.12), respectively. Similarly, the reconstruction algorithm is applied to these four output sequences to recover $\{c_{LL,k_1,k_2}^{(n)}\}$. The filter sequences are the decomposition sequences $\{a_k\}$, $\{b_k\}$ and two-scale sequences $\{p_k\}$, $\{q_k\}$ that govern the two-scale relations of $\phi(t)$ and $\psi(t)$ as in (1.3.5) and (1.3.7).

Algorithm 6.3 (two-dimensional decomposition algorithm).

$$
\begin{cases}
c_{LL,k_1,k_2}^{(n-1)} = \sum_{\ell_1,\ell_2} a_{\ell_1-2k_1} a_{\ell_2-2k_2} c_{LL,\ell_1,\ell_2}^{(n)}, \\
d_{LH,k_1,k_2}^{(n-1)} = \sum_{\ell_1,\ell_2} a_{\ell_1-2k_1} b_{\ell_2-2k_2} c_{LL,\ell_1,\ell_2}^{(n)}, \\
d_{HL,k_1,k_2}^{(n-1)} = \sum_{\ell_1,\ell_2} b_{\ell_1-2k_1} a_{\ell_2-2k_2} c_{LL,\ell_1,\ell_2}^{(n)}, \\
d_{HH,k_1,k_2}^{(n-1)} = \sum_{\ell_1,\ell_2} b_{\ell_1-2k_1} b_{\ell_2-2k_2} c_{LL,\ell_1,\ell_2}^{(n)}.
\end{cases}
$$

Algorithm 6.4 (two-dimensional reconstruction algorithm).

$$
\begin{aligned}
c_{LL,k_1,k_2}^{(n)} = \sum_{\ell_1,\ell_2} \{ & p_{k_1-2\ell_1} p_{k_2-2\ell_2} c_{LL,\ell_1,\ell_2}^{(n-1)} \\
& + p_{k_1-2\ell_1} q_{k_2-2\ell_2} d_{LH,\ell_1,\ell_2}^{(n-1)} + q_{k_1-2\ell_1} p_{k_2-2\ell_2} d_{HL,\ell_1,\ell_2}^{(n-1)} \\
& + q_{k_1-2\ell_1} q_{k_2-2\ell_2} d_{HH,\ell_1,\ell_2}^{(n-1)} \}.
\end{aligned}
$$

Recall that for orthogonal decomposition (i.e., $V_{n-1} \perp W_{n-1}$) the duality principle in Theorem 6.3 can be applied to interchange the pairs $(\{a_k\}, \{b_k\})$ and $(\{\bar{p}_{-k}\}, \{\bar{q}_{-k}\})$ by interchanging the original bases with their duals. In fact, as mentioned at the end of section 6.2, the same pair of filter sequences, say $(\{p_k\}, \{q_k\})$, can be used for both decomposition and reconstruction, when the change of bases is performed twice: the first time from $\phi_{n,k_1}(x)\phi_{n,k_2}(y)$ to its dual basis $\tilde{\phi}_{n,k_1}(x)\tilde{\phi}_{n,k_2}(y)$ before decomposition and the second time from $\tilde{\phi}_{n-1,k_1}(x)\tilde{\phi}_{n-1,k_2}(y)$, $\tilde{\phi}_{n-1,k_1}(x)\tilde{\psi}_{n-1,k_2}(y)$, $\tilde{\psi}_{n-1,k_1}(x)\tilde{\phi}_{n-1,k_2}(y)$, and $\tilde{\psi}_{n-1,k_1}(x)\tilde{\psi}_{n-1,k_2}(y)$ back to the original bases for the LL, LH, HL, and HH components before reconstruction.

In terms of frequency-domain interpretation, the four components $f_{n-1}(x,y)$, $g_{LH,n-1}(x,y)$, $g_{HL,n-1}(x,y)$, and $g_{HH,n-1}(x,y)$ of $f_n(x,y)$ in (6.4.11) give localized spatial information of $f_n(x,y)$ in the LL, LH, HL, and HH frequency ranges, respectively, as shown in Figure 6.2. Here, a square is used to represent the RMS bandwidth of $f_n(x,y)$ in both frequency directions, since we have chosen $n_1 = n_2 = n$.

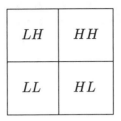

LH	*HH*
LL	*HL*

FIG. 6.2.   *Partition of RMS bandwidth into low- and high-frequency ranges.*

In general, returning to (6.4.8) and allowing n_1 to be different from n_2, we may write $f_{n_1,n_2}(x,y)$ in terms of the scaling function $\phi(2^{n_1}x - k_1)\phi(2^{n_2}y - k_2)$, namely,

$$f_{n_1,n_2}(x,y) = \sum_{k_1,k_2 \in \mathbb{Z}} c^{(n_1,n_2)}_{LL,k_1,k_2} \phi(2^{n_1}x - k_1)\phi(2^{n_2}y - k_2), \qquad (6.4.17)$$

and apply Algorithm 6.3 to $\{c^{(n_1,n_2)}_{LL,k_1,k_2}\}$ to yield $\{c^{(n_1-1,n_2-1)}_{LL,k_1,k_2}\}$, $\{d^{(n_1-1,n_2-1)}_{LH,k_1,k_2}\}$, $\{d^{(n_1-1,n_2-1)}_{HL,k_1,k_2}\}$, and $\{d^{(n_1-1,n_2-1)}_{HH,k_1k_2}\}$, keeping in mind that

$$f_{n_1,n_2}(x,y) = f_{n_1-1,n_2-1}(x,y) + g_{LH,n_1-1,n_2-1}(x,y)$$
$$+ g_{HL,n_1-1,n_2-1}(x,y) + g_{HH,n_1-1,n_2-1}(x,y)$$

with

$$\begin{cases} f_{n_1-1,n_2-1}(x,y) = \displaystyle\sum_{k_1,k_2 \in \mathbb{Z}} c^{(n_1-1,n_2-1)}_{LL,k_1,k_2} \phi(2^{n_1-1}x - k_1)\phi(2^{n_2-1}y - k_2), \\[2mm] g_{LH,n_1-1,n_2-1}(x,y) = \displaystyle\sum_{k_1,k_2 \in \mathbb{Z}} d^{(n_1-1,n_2-1)}_{LH,k_1,k_2} \phi(2^{n_1-1}x - k_1)\psi(2^{n_2-1}y - k_2), \\[2mm] g_{HL,n_1-1,n_2-1}(x,y) = \displaystyle\sum_{k_1,k_2 \in \mathbb{Z}} d^{(n_1-1,n_2-1)}_{HL,k_1,k_2} \psi(2^{n_1-1}x - k_1)\phi(2^{n_2-1}y - k_2), \\[2mm] g_{HH,n_1-1,n_2-1}(x,y) = \displaystyle\sum_{k_1,k_2 \in \mathbb{Z}} d^{(n_1-1,n_2-1)}_{HH,k_1,k_2} \psi(2^{n_1-1}x - k_1)\psi(2^{n_2-1}y - k_2). \end{cases}$$

The only difference in frequency-domain interpretation of this more general consideration is that the square in Figure 6.2 is allowed to become rectangular as shown in Figure 6.3 where $n_1 > n_2$.

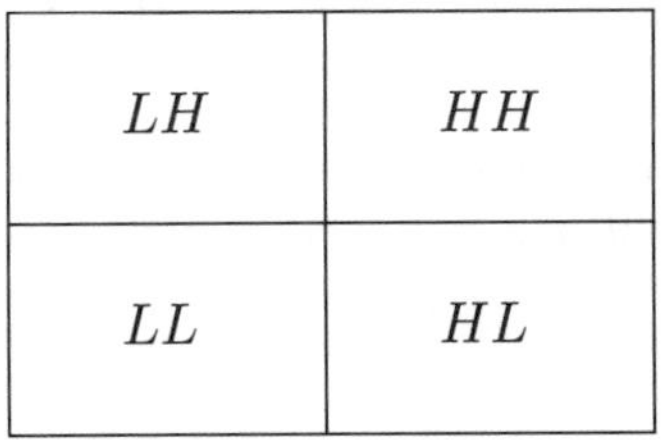

LH	*HH*
LL	*HL*

FIG. 6.3.   *Partition of individual RMS bandwidths.*

Now, the low-frequency portion of LH or HL (i.e., $g_{LH,n_1-1,n_2-1}(x,y)$ or $g_{HL,n_1-1,n_2-1}(x,y)$) can be further decomposed by using the one-dimensional decomposition algorithm as given in (1.3.11) (see also section 5.3, (5.3.19)). For example, we may decompose $g_{LH,n_1-1,n_2-1}(x,y)$ as

$$g_{LH,n_1-1,n_2-1}(x,y) = g_{LL,H,n_1-2,n_2-1}(x,y) + g_{LH,H,n_1-2,n_2-1}(x,y),$$

and these two components represent the octave bands LL,H and LH,H as shown in Figure 6.4. However, if a high-frequency direction such as H in HL is decomposed, we must interchange the filter sequences $\{a_k\}$ and $\{b_k\}$ to yield the L and H components of the H direction in HL. The result is a "wavelet packet" decomposition, and its frequency-domain interpretation is shown in Figure 6.5. Of course, both directions can be decomposed at the same time by applying the two-dimensional decomposition algorithm as shown in Figure 6.6, but the filter sequences $\{a_k\}$ and $\{b_k\}$ must be switched when decomposing high-frequency components, resulting in wavelet packet decompositions.

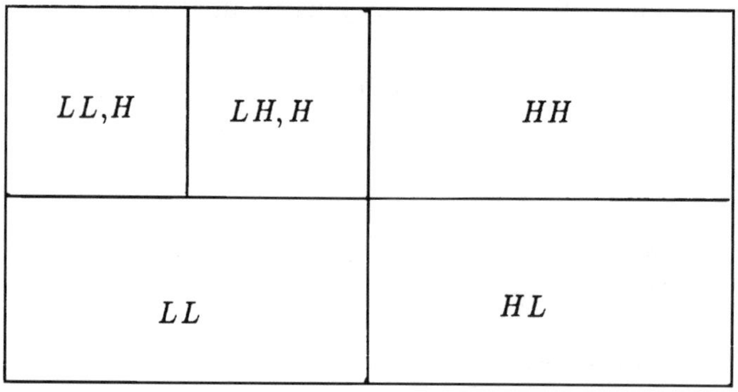

FIG. 6.4.   *Further partition of a low-frequency direction.*

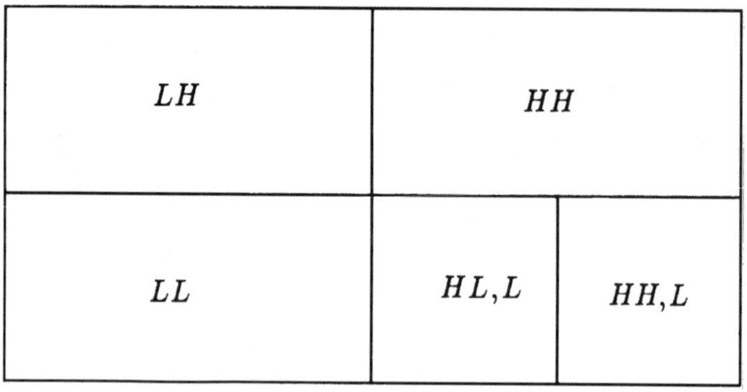

FIG. 6.5.   *Further partition of a high-frequency direction.*

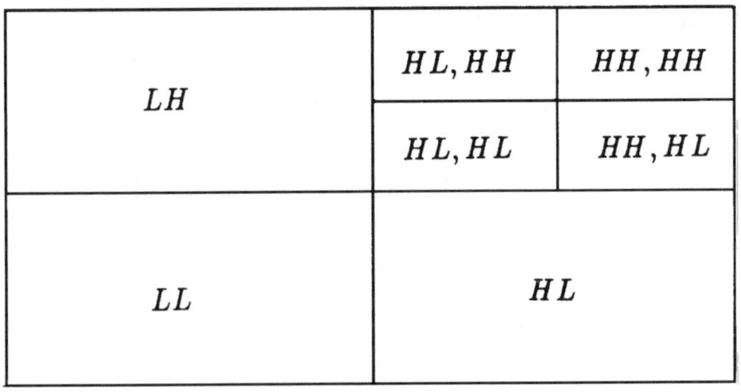

LH	HL, HH	HH, HH
	HL, HL	HH, HL

LL	HL

FIG. 6.6.   *The two-dimensional wavelet packet decomposition.*

6.5. The need for boundary wavelets.

The importance of the IWT with analyzing wavelet $\psi(t)$ is its ability to bring out the details of a signal $f(t)$ under investigation. For the sake of convenience in argument, assume that $\psi(t)$ has finite time duration (i.e., compact support), center at the origin, and m vanishing moments. Then at each point t_0 where $f(t)$ has a Taylor expansion

$$f(t) = \sum_{k=0}^{m-1} \frac{f^{(k)}(t_0)}{k!}(t - t_0)^k + R_m(t), \qquad |t - t_0| < \delta < 1, \qquad (6.5.1)$$

we have

$$(W_\psi f)(t_0, a) = \frac{1}{\sqrt{a}} \int_{|t-t_0|<\delta} R_m(t)\overline{\psi\left(\frac{t - t_0}{a}\right)} dt$$

when $0 < a \leq \delta$. Since $R_m(t) = O(\delta^m)$, we see that, at least for scale $a \leq \delta$, $(W_\psi f)(t_0, a)$ is very small. Therefore, at those points t_0 where (6.5.1) is not valid, or if (6.5.1) holds only for very small δ but $a > \delta$, then $(W_\psi f)(t_0, a)$ is relatively larger. This is the reason that the IWT can bring out the details of a signal $f(t)$ by varying the scale parameter $a > 0$.

Now suppose that $f(t)$ is defined only on a bounded interval $[c, d]$. Then for any t_0 that is very close to the endpoints c or d, the radius δ in (6.5.1) is very small and it takes only very small value of the scale a ($a \leq \delta$) to realize small values of $(W_\psi f)(t_0, a)$, even though $f(t)$ may be very smooth and has low frequency at t_0. Of course at the endpoints c and d themselves, the IWT values $(W_\psi f)(c, a)$ and $(W_\psi f)(d, a)$ are usually not small at all. Hence, it is necessary to introduce wavelets $\psi(t)$ that have vanishing moments relative to a bounded interval; namely,

$$\int_c^d t^\ell \psi(t) dt = 0, \qquad \ell = 0, \ldots, m - 1. \qquad (6.5.2)$$

Let us only consider the DWT with

$$\psi_{n,k}(t) := 2^{n/2} \psi(2^n t - k) \qquad (6.5.3)$$

and assume that the support of $\psi(t)$ is

$$\operatorname{supp} \psi = [0, N], \qquad (6.5.4)$$

where N is some positive integer. In addition, let the bounded interval of interest be

$$[c, d] = [0, 1]. \qquad (6.5.5)$$

Then since it follows from (6.5.3)–(6.5.4) that $\psi_{n,k}(t)$ vanishes for $t \notin \left[\frac{k}{2^n}, \frac{k+N}{2^n}\right]$, we have, for $2^n \geq N$,

$$\int_{-\infty}^{\infty} t^\ell \psi_{n,k}(t)dt = \int_0^1 t^\ell \psi_{n,k}(t)dt, \qquad 0 \leq k \leq 2^n - N. \qquad (6.5.6)$$

Hence, for the interval [0,1] in (6.5.5), we can keep the wavelets $\psi_{n,k}(t)$ for $0 \leq k \leq 2^n - N$ and only need new ones for $k = -N+1, \ldots, -1$ or $k = 2^n - N + 1, \ldots, 2^n - 1$. That is, we only need a total of $2(N-1)$ new wavelets, which will be called *boundary wavelets,* with $(N-1)$ for the boundary (or endpoint) 0 and $(N-1)$ for the boundary (or endpoint) 1. In fact, as we will see in section 6.7, the actual number is reduced by $\frac{1}{2}$ when orthogonal wavelet decomposition of an MRA for a bounded interval is considered. In any case, it is important to emphasize that the number of required boundary wavelets is independent of the resolution governed by 2^n. By a similar argument, we can also count the number of "boundary scaling functions" that must be constructed. This number again depends only on the size of the support of the given scaling function $\phi(t)$.

The general Hilbert space framework developed in section 6.2 can be followed for the construction and study of such boundary scaling functions and wavelets as well as their corresponding decomposition and reconstruction algorithms, when certain compactly supported semiorthogonal (and particularly orthonormal) scaling functions and wavelets for $L^2(-\infty, \infty)$ are given. More details will be given in the next two sections.

6.6. Spline functions on a bounded interval.

In order to prepare for our discussion of B-splines and B-wavelets with arbitrary knots in the next section, we introduce, in this section, the notion of normalized B-splines in terms of divided differences.

Definition 6.1 (divided differences). *Let*

$$\mathbf{t}: \ \cdots \leq t_0 \leq t_1 \leq t_2 \leq \cdots \qquad (6.6.1)$$

be a nondecreasing sequence of real numbers. Then the divided difference (dd) of a (sufficiently smooth) function $f(t)$ is defined as follows: the zeroth dd of $f(t)$ at $t = t_k$ is

$$[t_k]f := f(t_k);$$

the first dd of $f(t)$ at $\{t_k \leq t_{k+1}\}$ is

$$[t_k, t_{k+1}]f := \begin{cases} f'(t_k) & \text{if } t_k = t_{k+1}, \\ \frac{f(t_{k+1}) - f(t_k)}{t_{k+1} - t_k} & \text{if } t_k \neq t_{k+1}; \end{cases}$$

and, for any $m > 0$, the mth dd of $f(t)$ at $\{t_k \leq \cdots \leq t_{k+m}\}$ is

$$[t_k, \ldots, t_{k+m}]f := \begin{cases} \frac{f^{(m)}(t_k)}{m!} & \text{if } t_k = \cdots = t_{k+m}, \\ \frac{[t_{k+1}, \ldots, t_{k+m}]f - [t_k, \ldots, t_{k+m-1}]f}{t_{k+m} - t_k} & \text{if } t_k \neq t_{k+m}. \end{cases}$$

Observe that if $t_k \neq t_{k+m}$ but $t_k = \cdots = t_{k+m-1}$ in the numerator of the above definition, then by this same definition (with m replaced by $m-1$) we have $[t_k, \ldots, t_{k+m-1}]f = \frac{f^{(m-1)}(t_k)}{(m-1)!}$, etc.

We also remark that the above definition allows us to extend the definition to any finite set of real numbers without the ordering requirement (6.6.1). In other words, for any permutation $\{t'_k, \ldots, t'_{k+m}\}$ of $\{t_k, \ldots, t_{k+m}\}$, we may consider

$$[t'_k, \ldots, t'_{k+m}]f = [t_k, \ldots, t_{k+m}]f. \tag{6.6.2}$$

However, the nondecreasing order in (6.6.1) facilitates the computational procedure as given in the following.

Algorithm 6.5 (divided difference table).

t_k	$[t_k]f$		
		$[t_k, t_{k+1}]f$	
t_{k+1}	$[t_{k+1}]f$		
		$[t_{k+1}, t_{k+2}]f$	
t_{k+2}	$[t_{k+2}]f$		
$\vdots$	$\vdots$		$\cdots [t_k, \ldots, t_{k+m}]f$
t_{k+m-2}	$[t_{k+m-2}]f$		
		$[t_{k+m-2}, t_{k+m-1}]f$	
t_{k+m-1}	$[t_{k+m-1}]f$		
		$[t_{k+m-1}, t_{k+m}]f$	
t_{k+m}	$[t_{k+m}]f$		

Next, we introduce the notion of "truncated powers." For any real number x, we set

$$x_+ := \begin{cases} x & \text{if } x > 0, \\ 0 & \text{if } x \leq 0, \end{cases} \tag{6.6.3}$$

and, for any nonnegative integer n, we define

$$x_+^n := (x_+)^n = \begin{cases} x^n & \text{if } x > 0, \\ 0 & \text{if } x \leq 0. \end{cases} \tag{6.6.4}$$

Here, 0^0 is considered to be 0. Hence, for any fixed real number x and positive integer m, the function

$$f_x(t) := (x - t)_+^{m-1}, \tag{6.6.5}$$

as a function of the variable t, satisfies

$$\begin{cases} f_x(t) \in C^{m-2}, \\ f_x|_{[x,\infty)} \text{ and } f_x|_{(-\infty,x]} \text{ in } \pi_{m-1}, \end{cases}$$

meaning that $f_x(t)$ is $(m-2)$-times continuously differentiable everywhere, and the restrictions of $f_x(t)$ on both intervals $[x, \infty)$ and $(-\infty, x]$ are polynomials in t of order m (or degree $m - 1$; see section 3.4). We will call $f_x(t)$ in (6.6.5) a spline function of order m with a single "knot" at $t = x$. In this section, since we are only concerned with spline functions on a bounded interval, we need the notation

$$C^k[c, d] = \{f(t): f(t), \ldots, f^{(k)}(t) \text{ exist and are continuous on } [c, d]\}.$$

The difference between this notation and the notation C^k introduced in section 3.4 is that the derivatives of the functions $f(t)$ in $C^k[c, d]$ are taken only relative to the interval $[c, d]$; i.e., only one-sided limits at c and d are used in the definition of derivatives.

Definition 6.2 (splines on a bounded interval). *Let m and n be positive integers and*

$$\mathbf{x}: \; c = x_0 < x_1 < \cdots < x_n = d. \tag{6.6.6}$$

Then the functions in the collection

$$S_{\mathbf{x},m} := \{f(t) \in C^{m-2}[c, d]: f|_{[x_{i-1}, x_i]} \in \pi_{m-1}, \quad i = 1, \ldots, n\} \tag{6.6.7}$$

are called spline functions of order m with knot sequence $\mathbf{x}$ on $[c, d]$.

It is clear that $S_{\mathbf{x},m}$ is a vector space. For the example $f_x(t)$ in (6.6.5), if x is chosen to be one of the knots, say x_k, $0 \le k \le n$ in (6.6.6), then $f_{x_k}(t) \in S_{\mathbf{x},m}$. Hence, any linear combination

$$\sum_{k=0}^{n} c_k f_{x_k}(t), \quad c_k \text{ arbitrary constants}, \tag{6.6.8}$$

is in $S_{\mathbf{x},m}$. In particular, by using the notation

$$[x_k, \ldots, x_{k+m}]_x f(x, t) \tag{6.6.9}$$

to mean that the mth divided difference is taken at $x = x_k, \ldots, x_{k+m}$ while t is kept fixed, we see that

$$[x_k, \ldots, x_{k+m}]_x f_x(t) = [x_k, \ldots, x_{k+m}]_x (x - t)_+^{m-1}$$

can be written in the form of (6.6.8) for some constants c_k and, hence, is an mth-order spline function with knots at $x_k < \cdots < x_{k+m}$.

Next, let us extend the definition of the knot sequence $\mathbf{x}$ in (6.6.6) to

$$\mathbf{x}: \; c = x_{-m+1} = \cdots = x_0 < x_1 < \cdots < x_n = \cdots = x_{n+m-1} = d \tag{6.6.10}$$

simply by giving more labels to the endpoints c and d. In doing so, we can use the notation (6.6.9) to apply the definition of divided differences with possible coincident points as $\mathbf{t}$ in (6.6.1).

Definition 6.3. *For each $k = -m+1, \ldots, n-1$, let*

$$N_{\mathbf{x},m;k}(t) := (x_{k+m} - x_k)[x_k, \ldots, x_{k+m}]_x (x - t)_+^{m-1} \qquad (6.6.11)$$

and call it the kth (normalized) B-spline of order m with knot sequence $\mathbf{x}$ given in (6.6.10).

We have seen that if $0 \le k \le n - m$, then $N_{\mathbf{x},m;k}(t)$ can be written as (6.6.8) and, hence, is in $S_{\mathbf{x},m}$. On the other hand, for $-m+1 \le k \le -1$ or $n-m+1 \le k \le n-1$, we see from Definition 6.1 that certain derivatives of the truncated powers given by (6.6.5) are taken with respect to the variable x at $x = c$ or d only. Since this operation does not change the order of smoothness at the "interior knots" $x_1, \ldots, x_{n-1}$, the functions $N_{\mathbf{x},m;k}(t)$ are still in $C^{m-2}[c, d]$. Hence, all of the normalized B-splines $N_{\mathbf{x},m;k}(t)$, $k = -m+1, \ldots, n-1$ are in $S_{\mathbf{x},m}$. In fact, they constitute a basis of this vector space.

Theorem 6.4. *The collection of mth-order B-splines $N_{\mathbf{x},m;k}(t)$, $k = -m+ 1, \ldots, n - 1$ with knot sequence $\mathbf{x}$ given by (6.6.10) is a basis of the spline space $S_{\mathbf{x},m}$ defined in (6.6.7). Furthermore, these B-splines have the following properties:*

(i) $\operatorname{supp} N_{\mathbf{x},m;k} = [c, d] \cap [x_k, x_{k+m}]$;
(ii) $N_{\mathbf{x},m;k}(t) > 0$ *for t in the interior of* $\operatorname{supp} N_{\mathbf{x},m;k}$;
(iii) $\sum_{k=-m+1}^{n-1} N_{\mathbf{x},m;k}(t) = 1$ *for all $t \in [c, d]$*;
(iv) *for $m \ge 2$,*

$$N'_{\mathbf{x},m;k}(t) = \frac{m-1}{x_{k+m-1} - x_k} N_{\mathbf{x},m-1;k}(t) - \frac{m-1}{x_{k+m} - x_{k+1}} N_{\mathbf{x},m-1;k+1}(t);$$

(v) $N_{\mathbf{x},m;k}(t)$ *can be computed from $N_{\mathbf{x},m-1,k}(t)$ by using the identity*

$$N_{\mathbf{x},m;k}(t) = \frac{t - x_k}{x_{k+m-1} - x_k} N_{\mathbf{x},m-1;k}(t) + \frac{x_{k+m} - t}{x_{k+m} - x_{k+1}} N_{\mathbf{x},m-1;k+1}(t),$$
$$(6.6.12a)$$

where

$$N_{\mathbf{x},1;k}(t) = \chi_{[x_k, x_{k+1})}(t), \qquad k = 0, \ldots, n - 1. \qquad (6.6.12b)$$

For simplicity, let us first restrict our attention to integer knots. That is, we consider the bounded interval

$$[c, d] = [0, n], \qquad (6.6.13)$$

where n is some positive integer, and

$$x_k = k, \qquad k = 0, \ldots, n \qquad (6.6.14)$$

for the knot sequence $\mathbf{x}$ in (6.6.10).

Example 6.10. Let $N_m(t)$ be the mth-order cardinal B-spline introduced in section 3.4 of Chapter 3, and consider the mth-order B-splines $N_{m,\mathbf{x};k}(t)$ in (6.6.11) for the bounded interval (6.6.13) with knot sequence $\mathbf{x}$ given by (6.6.14). Then by the support property (i) in Theorem 6.4, it can be easily shown that the B-splines $N_{m,\mathbf{x};k}(t)$ are simply the cardinal B-splines $N_m(t-k)$ when k is so chosen that supp $N_m(t-k)$ lies in $[0,n]$; i.e., we have

$$N_{m,\mathbf{x};k}(t) \equiv N_m(t-k), \qquad k = 0, \ldots, n-m. \qquad (6.6.15)$$

These are the "interior B-splines" for the bounded interval $[0,n]$ (6.4.13). The remaining B-splines

$$\begin{cases} N_{m,\mathbf{x};k}(t), & k = -m+1, \ldots, -1, \text{ and} \\ N_{m,\mathbf{x};k}(t), & k = n-m+1, \ldots, n-1 \end{cases} \qquad (6.6.16)$$

are the "boundary B-splines" for the interval $[0,n]$. Here, the first group (i.e., $k = -m+1, \ldots, -1$) is for the boundary (or endpoint) $c = 0$, while the second group (i.e., $k = n-m+1, \ldots, n-1$) is for the boundary $d = n$.

In the following, we demonstrate how to compute these boundary B-splines. Since the boundary B-splines at $c = 0$ are symmetric reflections of those at $c = n$ (simply by replacing t with $n-t$), it is sufficient to construct the first half in (6.6.16).

Two methods have already been introduced in this section: one is to apply Algorithm 6.5 directly to the truncated powers as in the definition of $N_{m,\mathbf{x};k}(t)$ in (6.6.11) and the other is to follow the recursive scheme given in (6.6.12a) with initial function (6.6.12b) in Theorem 6.4(v). Both algorithms are quite efficient, but we will only discuss the second method.

Example 6.11. For $m = 2$ (i.e., linear splines), there is only one boundary B-spline $N_{\mathbf{x},2;-1}(t)$ at the boundary $c = 0$. Since $k = -1$ and $N_{\mathbf{x},1,-1}(t) \equiv 0$, the first term in (6.6.12a) vanishes, so that by (6.6.12a)–(6.6.12b) together we have

$$N_{\mathbf{x},2;-1}(t) = \frac{1-t}{1-0} N_{\mathbf{x},1;0}(t) = (1-t)\chi_{[0,1)}(t).$$

This is simply the restriction of $N_2(t+1)$ on the interval $[0,1)$.

Example 6.12. For $m = 3$ (i.e., quadratic splines), there are two boundary B-splines $N_{\mathbf{x},3;-2}(t)$ and $N_{\mathbf{x},3;-1}(t)$ at the boundary $c = 0$. For the first one, since $N_{\mathbf{x},2;-2}(t) \equiv 0$, the first term in (6.6.12a) vanishes. Hence, by (6.6.12a) and using the result in Example 6.11, we have

$$N_{\mathbf{x},3;-2}(t) = \frac{1-t}{1-0} N_{\mathbf{x},2;-1}(t) = (1-t)^2 \chi_{[0,1)}(t).$$

To compute the second boundary B-spline, we apply (6.6.12a) and Examples 6.10–6.11 to obtain

$$N_{\mathbf{x},3;-1}(t) = \frac{t-0}{1-0} N_{\mathbf{x},2;-1}(t) + \frac{2-t}{2-0} N_{\mathbf{x},2;0}(t)$$

$$= t N_{\mathbf{x},2;-1}(t) + \frac{2-t}{2} N_2(t)$$

$$= \begin{cases} 2t - \frac{3}{2}t^2 & \text{if } 0 \le t < 1, \\ \frac{1}{2}(2-t)^2 & \text{if } 1 \le t < 2, \\ 0 & \text{otherwise.} \end{cases}$$

Example 6.13. For $m = 4$ (i.e., cubic splines), there are three boundary B-splines $N_{\mathbf{x},4;-3}(t)$, $N_{\mathbf{x},4;-2}(t)$, and $N_{\mathbf{x},4;-1}(t)$ at the boundary $c = 0$. Again for the first one, since $N_{\mathbf{x},3;-3}(t) \equiv 0$, it follows from (6.6.12a) and Example 6.12 that

$$N_{\mathbf{x},4;-3}(t) = \frac{1-t}{1-0} N_{\mathbf{x},3;-2}(t) = (1-t)^3 \chi_{[0,1)}(t).$$

For $k = -2$ in (6.6.12a), we again use the result in Example 6.12 to obtain

$$N_{\mathbf{x},4;-2}(t) = \frac{t-0}{1-0} N_{\mathbf{x},3;-2}(t) + \frac{2-t}{2-0} N_{\mathbf{x},3;-1}(t)$$

$$= \begin{cases} 3t - \frac{9}{2}t^2 + \frac{7}{4}t^3 & \text{if } 0 \le t < 1, \\ \frac{1}{4}(2-t)^3 & \text{if } 1 \le t < 2, \\ 0 & \text{otherwise.} \end{cases}$$

Finally, to calculate the third boundary cubic B-spline $N_{\mathbf{x},4;-1}(t)$, we apply (6.6.12a) and Examples 6.10 and 6.12 to obtain

$$N_{\mathbf{x},4;-1}(t) = \frac{t-0}{2-0} N_{\mathbf{x},3;-1}(t) + \frac{3-t}{3-0} N_{\mathbf{x},3;0}(t)$$

$$= \frac{t}{2} N_{\mathbf{x},3;-1}(t) + \frac{3-t}{3} N_4(t)$$

$$= \begin{cases} \frac{3}{2}t^2 - \frac{11}{12}t^3 & \text{if } 0 \le t < 1, \\ \frac{7}{12} + \frac{1}{4}(t-1) - \frac{5}{4}(t-1)^2 + \frac{7}{12}(t-1)^3 & \text{if } 1 \le t < 2, \\ \frac{1}{6}(3-t)^3 & \text{if } 2 \le t < 3, \\ 0 & \text{otherwise.} \end{cases}$$

6.7. Boundary spline wavelets with arbitrary knots.

When the DWT was introduced through an MRA $\{V_n\}$ of L^2, a two-scale relation for the scaling function $\phi(t)$ is used to govern the subspace property

$V_{n-1} \subset V_n$. However, this relation is no longer valid when we restrict our attention to a bounded interval. Fortunately, the "interior scaling functions" are still $\phi(2^{n-1}t - k)$ and hence remain to enjoy the same two-scale relation as in the (unbounded) $L^2(-\infty, \infty)$ setting, and only a few "boundary scaling functions" are needed to generate the spaces V_n. It is important to remember that this number (of boundary scaling functions) remains unchanged as $n \to \infty$, and once the boundary scaling functions for the "lowest" level, say V_{n_0}, have been constructed, those for the higher levels V_n, where $n > m$, are simply their scaled versions (by 2^{-n+n_0}).

In the above section, we have demonstrated, in Examples 6.11–6.13, how the recursive algorithm (6.6.12a)–(6.6.12b) in Theorem 6.4 is applied to compute all the boundary scaling functions (or B-splines of an arbitrary order m) in the spline setting. Hence, for each (fixed) order m, we have all the scaling functions as *basis functions* for each level of the nested sequence of subspaces $\{V_n\}$. (We remark that the notion of MRA for a bounded interval has to be modified. For example, we no longer have $n \to -\infty$ but rather a "ground level.")

Our strategy is to follow the general Hilbert space framework introduced in section 6.2: first, find the matrix P that relates the basis (i.e., both boundary and interior scaling functions) of V_{n-1} to the corresponding (scaled) basis of V_n; then, find the basis (i.e., both boundary and interior wavelets) of the orthogonal complementary subspace W_{n-1} of V_n relative to V_{n-1}, as well as the matrix Q; and, finally, by applying these two matrices P, Q and the Gramian matrices G, H, K of the three bases, compute the decomposition matrices A and B as well as the dual scaling functions and wavelets by applying Theorem 6.2 and (6.2.32), respectively.

To facilitate our presentation, we only consider the (polynomial) spline setting with exactly one additional knot in-between two existing ones when we go up the "MRA ladder." That is, we consider two knot sequences $\mathbf{t}$ and $\mathbf{x}$, given by

$$\begin{cases} \mathbf{t}: \ c = t_{-m+1} = \cdots = t_0 < \cdots < t_{2n} = \cdots = t_{2n+m-1} = d, \\ \mathbf{x}: \ c = x_{-m+1} = \cdots = x_0 < \cdots < x_n = \cdots = x_{n+m-1} = d, \end{cases} \qquad (6.7.1)$$

with

$$t_{2i} = x_i, \qquad i = 0, \ldots, n, \qquad (6.7.2)$$

where m and n are any positive integers. Note that $\mathbf{x}$ is the same as the one in (6.6.10) and $\mathbf{t}$ is a refinement of $\mathbf{x}$ by inserting a knot t_{2i+1} between x_i and x_{i+1} for each $i = 0, \ldots, n - 1$. It is clear from Definition 6.2 that the spline space $S_{\mathbf{x},m}$ of order m with knot sequence $\mathbf{x}$ is a proper subspace of the spline space $S_{\mathbf{t},m}$ of the same order but with a finer knot sequence. Hence, by Theorem 6.4 and following the notations used in section 6.2, we have

$$\begin{cases} U = S_{\mathbf{t},m} \quad \text{with basis} \\ u_i = N_{\mathbf{t},m;i-m}(t), \qquad i = 1, \ldots, 2n + m - 1, \end{cases} \qquad (6.7.3)$$

and

$$\begin{cases} V = S_{\mathbf{x},m} \quad \text{with basis} \\ v_i = N_{\mathbf{x},m;i-m}(t), \qquad i = 1, \dots, n+m-1, \end{cases} \qquad (6.7.4)$$

that satisfy

$$\begin{cases} V \subset U, \\ \dim U = 2n + m - 1 \text{ and } \dim V = n + m - 1. \end{cases}$$

So, the orthogonal complementary subspace $W = W_{\mathbf{x},\mathbf{t},m}$ of U relative to V has dimension n (independent of the order m of the spline space), and the basis functions

$$w_i = \psi_{\mathbf{x},\mathbf{t},m;i-m}(t), \qquad i = 1, \dots, n \qquad (6.7.5)$$

of W that have minimum support and appropriately normalized (such as $\|w_i\| = 1$ for all i) will be called *mth-order B-wavelets with knot sequence* **t** *relative to knot sequence* **x**.

Let us first give an efficient algorithm for computing the matrix

$$P = P_m = \left[p_{m;i,j} \right]_{\substack{1 \le i \le n+m-1 \\ 1 \le j \le 2n+m-1}} \qquad (6.7.6)$$

that relates the B-spline bases in (6.7.3)–(6.7.4) in the form of

$$N_{\mathbf{x},m;i-m}(t) = \sum_{j=1}^{2n+m-1} p_{m;i,j} N_{\mathbf{t},m;j-m}(t), \qquad i = 1, \dots, n+m-1. \quad (6.7.7)$$

Algorithm 6.6 (computation of B-spline two-scale matrices).
The initial matrix (for $m = 1$) is given by an $n \times (2n)$ matrix

$$P_1 = \begin{bmatrix} 1 & 1 & 0 & 0 & 0 & 0 & \dots & 0 & 0 \\ 0 & 0 & 1 & 1 & 0 & 0 & \dots & 0 & 0 \\ \dots\dots\dots\dots\dots\dots\dots\dots\dots\dots\dots \\ 0 & 0 & \dots & 0 & 0 & 1 & 1 & 0 & 0 \\ 0 & 0 & \dots & \dots & \dots & 0 & 0 & 1 & 1 \end{bmatrix}. \qquad (6.7.8)$$

For each $m \ge 2$, we have

$$P_m = \begin{bmatrix} 1 & \frac{x_1 - t_1}{x_1 - x_0} & 0 & \dots & \dots & & \dots & & 0 \\ 0 & & & & & & & & \\ \vdots & & & \widehat{P}_m & & & & & \\ 0 & 0 & 0 & \dots & \dots & \frac{x_{n-1} - t_{2n-1}}{x_n - x_n} & & 1 & \end{bmatrix}, \qquad (6.7.9)$$

where

$$\widehat{P}_m = \left[\widehat{p}_{m;i,j} \right]_{\substack{1 \le i \le n+m-3 \\ 1 \le j \le 2n+m-2}} \qquad (6.7.10)$$

is an $(n + m - 3) \times (2n + m - 2)$ submatrix of P_m that can be computed recursively as follows:

$$\hat{p}_{m;i,j} = \frac{t_j - x_{i-m+1}}{x_i - x_{i-m+1}} p_{m-1;i,j} + \frac{x_{i+1} - t_j}{x_{i+1} - x_{i-m+2}} p_{m-1;i+1,j}, \qquad (6.7.11)$$

where $i = 1, \ldots, n + m - 3$ and $j = 1, \ldots, 2n + m - 2$.

Example 6.14. For the special case

$$\begin{cases} t_i = \dfrac{i}{2}, \text{ if } i = 0, \ldots, 2n, \text{ and} \\ t_{-m+1} = \cdots = t_{-1} = 0, t_{2n+1} = \cdots = t_{2n+m-1} = n, \end{cases}$$

we have $[c, d] = [0, n]$ and

$$\begin{cases} x_i = i, \text{ if } i = 0, \ldots, n, \text{ and} \\ x_{-m+1} = \cdots = x_{-1} = 0, x_{n+1} = \cdots = x_{n+m-1} = n. \end{cases}$$

The two-scale matrix P_1 is the same as (6.7.8). By applying (6.7.11) with $m = 2, 3, 4$, we have

$$P_2 = \frac{1}{2} \begin{bmatrix} 2 & 1 & 0 & 0 & 0 & & & & \\ 0 & 1 & 2 & 1 & 0 & & & \bigcirc & \\ 0 & 0 & 0 & 1 & 2 & 1 & & & \\ & & & & & \ddots & & & \\ & & \bigcirc & & & 1 & 2 & 1 & 0 \\ & & & & & 0 & 0 & 1 & 2 \end{bmatrix}_{(n+1) \times (2n+1)},$$

$$P_3 = \frac{1}{2^2} \begin{bmatrix} 4 & 2 & 0 & 0 & 0 & 0 & 0 & 0 \\ 0 & 2 & 3 & 1 & 0 & 0 & 0 & 0 \\ 0 & 0 & 1 & 3 & 3 & 1 & 0 & 0 \\ & & & 1 & 3 & 3 & 1 & \\ & & & & & \ddots & & \\ & & \bigcirc & & 1 & 3 & 3 & 1 & 0 & 0 \\ & & & & 0 & 0 & 1 & 3 & 2 & 0 \\ & & & & 0 & 0 & 0 & 0 & 2 & 4 \end{bmatrix}_{(n+2) \times (2n+2)},$$

$$P_4 = \frac{1}{2^3} \begin{bmatrix} 8 & 4 & 0 & 0 & 0 & 0 & 0 & 0 & 0 & 0 \\ 0 & 4 & 6 & \frac{3}{2} & 0 & 0 & 0 & 0 & 0 & 0 \\ 0 & 0 & 2 & \frac{11}{2} & 4 & 1 & 0 & 0 & 0 & 0 \\ 0 & 0 & 0 & 1 & 4 & 6 & 4 & 1 & 0 & 0 \\ 0 & 0 & 0 & 0 & 0 & 1 & 4 & 6 & 4 & 1 \\ & & & & & & \ddots & & & \\ & & \bigcirc & & & 1 & 4 & 6 & 4 & 1 & 0 & 0 & 0 \\ & & & & & 0 & 0 & 1 & 4 & \frac{11}{2} & 2 & 0 & 0 \\ & & & & & 0 & 0 & 0 & 0 & \frac{3}{2} & 6 & 4 & 0 \\ & & & & & 0 & 0 & 0 & 0 & 0 & 0 & 4 & 8 \end{bmatrix}_{(n+3) \times (2n+3)}.$$

To construct the B-wavelets $\psi_{\mathbf{x},\mathbf{t},m;i}(t)$ and determine their corresponding two-scale matrix $Q = Q_m$, we observe that these basis functions of W are linear combinations of the B-splines $N_{\mathbf{t},m;k}(t)$ and that they are orthogonal to the B-splines $N_{\mathbf{x},m;k}(t)$. Hence, the $(n+m) \times (2n+m-1)$ "master" matrix

$$R = \begin{bmatrix} N_{\mathbf{t},m;1-m}(t) & N_{\mathbf{t},m;2-m}(t) & \cdots & N_{\mathbf{t},m;2n-1}(t) \\ \langle N_{\mathbf{t},m;1-m}, N_{\mathbf{x},m;1-m} \rangle & \langle N_{\mathbf{t},m;2-m}, N_{\mathbf{x},m,1-m} \rangle & \cdots & \langle N_{\mathbf{t},m;2n-1}, N_{\mathbf{x},m;1-m} \rangle \\ \cdots\cdots\cdots\cdots\cdots\cdots\cdots\cdots\cdots\cdots\cdots\cdots\cdots\cdots\cdots\cdots\cdots\cdots\cdots\cdots\cdots\cdots \\ \langle N_{\mathbf{t},m;1-m}, N_{\mathbf{x},m;n-1} \rangle & \langle N_{\mathbf{t},m;2-m}, N_{\mathbf{x},m,n-1} \rangle & \cdots & \langle N_{\mathbf{t},m;2n-1}, N_{\mathbf{x},m;n-1} \rangle \end{bmatrix}$$
$$(6.7.12)$$

whose entries, with the exception of those in the first row, are constants, plays an important role. First, the determinants of the square submatrices $R_1, \ldots, R_n$ of dimension $n+m$ (consisting of $n+m$ consecutive entries of the first row) are linear combinations of the B-splines $N_{\mathbf{t},m,k}(t)$. Second, the inner product of $\det R_i$, $i = 1, \ldots, n$ with any of the B-splines $N_{\mathbf{x},m;k}(t)$ must vanish. The reason for the second conclusion is that $\langle \det R_i, N_{\mathbf{x},m;k} \rangle$ is the determinant of a matrix with two identical rows. Hence, $\det R_1, \ldots, \det R_n$, which are functions of t, seem to be ideal candidates for the n B-wavelets. Unfortunately, this approach fails, since these determinants are identically zero when n is sufficiently large. So, we must look for square submatrices of the master matrix of dimension independent of n. (Recall that $2n - 1$ is the number of knots in $\mathbf{t}$.) To do so, we observe that the two-scale relation of the mth-order cardinal B-wavelet $\psi_m(t)$ in (5.2.25) can be written as

$$\psi_m(t - (i - m)) = \sum_{k=0}^{3m-2} q_k N_m(2t - (k + 2i - 2m))$$

$$= \sum_{j=2i-m}^{2i+2m-2} q_{j-2i+m} N_m(2t - (j - m)).$$

This suggests the following formulation of the B-wavelets:

$$w_i = \psi_{\mathbf{x},\mathbf{t};m,i-m}(t) = \sum_{j=2i-m}^{2i+2m-2} q_{i,j} N_{\mathbf{t},m;j-m}(t)$$

$$= \sum_{j=2i-m}^{2i+2m-2} q_{ij} u_j,$$

for some constants $q_{i,j}$'s. This formulation of the B-wavelets w_i is reasonable only for the "interior" ones (i.e., w_i for $i = m, \ldots, n-m+1$). For the boundary wavelets (i.e., w_i for $i = 1, \ldots, m-1$ or $i = n-m+2, \ldots, n$), however, since the two (northwest and southeast) corners of the two-scale matrix P_m are not shifts by two entries as are the "interior" entries, we need to be much more careful. Indeed, for $i = 1, \ldots, m - 1$, for example, since the B-splines have two-scale relations of the form

$$v_i := N_{\mathbf{x},m;i-m}(t) = \sum_{j=i}^{2i} p_{i,j} u_j,$$

and the lengths of the $q_{i,j}$ sequences (for each fixed i) should have $2m - 2$ more terms than the $p_{i,j}$ sequences (for the same fixed i), it is reasonable to formulate the boundary wavelets w_i, $i = 1, \ldots, m - 1$, as linear combinations of u_j, for $j = i, \ldots, 2i + 2m - 2$. Of course, the other boundary wavelets w_i, $i = n - m + 2, \ldots, n$, must be formulated in a similar way. In summary, our proposed B-wavelets w_i, $i = 1, \ldots, n$, take on the following form.

(i) For $i = 1, \ldots, m - 1$ (i.e., left boundary),

$$w_i = \sum_{j=i}^{2i+2m-2} q_{i,j} u_j;$$

(ii) for $i = m, \ldots, n - m + 1$ (i.e., interior),

$$w_i = \sum_{j=2i-m}^{2i+2m-2} q_{i,j} u_j; \quad \text{and}$$

(iii) for $i = n - m + 2, \ldots, n$ (i.e., right boundary),

$$w_i = \sum_{j=2i-m}^{n+i+m-1} q_{i,j} u_j.$$

In other words we look for a two-scale matrix

$$Q_m = [q_{m;i,j}] = [q_{i,j}]_{\substack{1 \le i \le n \\ 1 \le j \le 2n+m-1}}, \tag{6.7.13}$$

where the $q_{i,j}$'s vanish for the following indices (i, j).

(α) For $i = 1, \ldots, m - 1$,

$$q_{i,j} = 0 \text{ if } 1 \le j \le i - 1 \text{ or } 2i + 2m - 1 \le j \le 2n + m - 1;$$

(β) for $i = m, \ldots, n - m + 1$,

$$q_{i,j} = 0 \text{ if } 1 \le j \le 2i - m - 1 \text{ or } 2i + 2m - 1 \le j \le 2n + m - 1;$$

(γ) for $i = n - m + 2, \ldots, n$,

$$q_{i,j} = 0 \text{ if } 1 \le j \le 2i - m - 1 \text{ or } n + i + m \le j \le 2n + m - 1.$$

From the formulation of the proposed w_i, $i = 1, \ldots, n$, in (i)–(iii) above, it follows that

$$\text{supp } w_i = [x_{i-m}, x_{i+m-1}]. \tag{6.7.14}$$

On the other hand, the support of $v_j = N_{\mathbf{x},m;j-m}(t)$ is given by

$$\text{supp } v_j = [x_{j-m}, x_j]. \tag{6.7.15}$$

As a consequence of (6.7.14)–(6.7.15), we already have the following orthogonality properties.

(a) For $i = 1, \ldots, m-1$,

$$\langle w_i, v_j \rangle = 0 \text{ if } i + 2m - 1 \leq j \leq n + m - 1;$$

(b) for $i = m, \ldots, n - m + 1$,

$$\langle w_i, v_j \rangle = 0 \text{ if } 1 \leq j \leq i - m \text{ or } i + 2m - 1 \leq j \leq n + m - 1;$$

(c) for $i = n - m + 2, \ldots, n$,

$$\langle w_i, v_j \rangle = 0 \text{ if } 1 \leq j \leq i - m.$$

Hence, to ensure that $w_i \perp V$, we only require the following orthogonality conditions.

(d) For $i = 1, \ldots, m-1$, we need

$$\langle w_i, v_j \rangle = 0 \text{ if } 1 \leq j \leq i + 2m - 2;$$

(e) for $i = m, \ldots, n - m + 1$, we need

$$\langle w_i, v_j \rangle = 0 \text{ if } i - m + 1 \leq j \leq i + 2m - 2;$$

(f) for $i = n - m + 2, \ldots, n$, we need

$$\langle w_i, v_j \rangle = 0 \text{ if } i - m + 1 \leq j \leq n + m - 1.$$

In view of the nature of the master matrix R in (6.7.12), the orthogonality conditions (d)–(f) are clearly satisfied by the following spline functions.

(i') For $i = 1, \ldots, m-1$,

$$\widehat{w}_i := \det \begin{bmatrix} u_i & u_{i+1} & \cdots & u_{2i+2m-2} \\ \langle u_i, v_1 \rangle & \langle u_{i+1}, v_1 \rangle & \cdots & \langle u_{2i+2m-2}, v_1 \rangle \\ \cdots\cdots\cdots\cdots\cdots\cdots\cdots\cdots\cdots\cdots\cdots\cdots\cdots\cdots\cdots \\ \langle u_i, v_{i+2m-2} \rangle & \langle u_{i+1}, v_{i+2m-2} \rangle & \cdots & \langle u_{2i+2m-2}, v_{i+2m-2} \rangle \end{bmatrix};$$

(ii') for $i = m, \ldots, n - m + 1$,

$$\widehat{w}_i := \det \begin{bmatrix} u_{2i-m} & u_{2i-m+1} & \cdots & u_{2i+2m-2} \\ \langle u_{2i-m}, v_{i-m+1} \rangle & \langle u_{2i-m+1}, v_{i-m+1} \rangle & \cdots & \langle u_{2i+2m-2}, v_{i-m+1} \rangle \\ \cdots\cdots\cdots\cdots\cdots\cdots\cdots\cdots\cdots\cdots\cdots\cdots\cdots\cdots\cdots \\ \langle u_{2i-m}, v_{i+2m-2} \rangle & \langle u_{2i-m+1}, v_{i+2m-2} \rangle & \cdots & \langle u_{2i+2m-2}, v_{i+2m-2} \rangle \end{bmatrix};$$

(iii') for $i = n - m + 2, \ldots, n$,

$$\widehat{w}_i := \det \begin{bmatrix} u_{2i-m} & u_{2i-m+1} & \cdots & u_{n+i+m-1} \\ \langle u_{2i-m}, v_{i-m+1} \rangle & \langle u_{2i-m+1}, v_{i-m+1} \rangle & \cdots & \langle u_{n+i+m-1}, v_{i-m+1} \rangle \\ \cdots\cdots\cdots\cdots\cdots\cdots\cdots\cdots\cdots\cdots\cdots\cdots\cdots\cdots\cdots\cdots\cdots \\ \langle u_{2i-m}, v_{n+m-1} \rangle & \langle u_{2i-m+1}, v_{n+m-1} \rangle & \cdots & \langle u_{n+i+m-1}, v_{n+m-1} \rangle \end{bmatrix}.$$

In addition, by expanding each of the determinants with respect to the first row, we see that each $\widehat{w}_i$ has the same form as w_i in (i)–(iii), $i = 1, \ldots, n$. Hence, we simply set

$$w_i = c_i \widehat{w}_i, \qquad i = 1, \ldots, n \tag{6.7.16}$$

for some normalization constants $c_i \neq 0$. One choice of c_i is to ensure $\|w_i\| = 1$. For uniform knots as in Example 6.14, we prefer to select c_i so that the interior rows of Q_m agree with the two-scale sequence for the $L^2(-\infty, \infty)$ setting; i.e.,

$$\begin{cases} q_{m;i,j} = q_{i,j} = q_{j-2i}, \\ \text{with } q_k = \dfrac{(-1)^k}{2^{m-1}} \displaystyle\sum_{\ell=0}^{m} \binom{m}{\ell} N_{2m}(k - \ell + 1) \end{cases} \tag{6.7.17}$$

(see (5.3.36)).

Regardless of the normalization constants, the choice of w_i in (6.7.16) gives rise to the matrix structure (α)–(γ) and ensures the orthogonality properties (a)–(f). Hence, $w_i \perp V$ for all $i = 1, \ldots, n$. It can also be shown that

$$\{w_1, \ldots, w_n\}$$

is a linearly independent set and, hence, is a basis of the wavelet space W. The other (nonzero) entries of the two-scale matrix Q_m can be computed by using (i')–(iii'). In addition, since all the bases $\{u_i\}, \{v_i\}, \{w_i\}$ of U, V, W, respectively (see (6.7.3)–(6.7.5) for the notations), are known, we can compute their Gramian matrices G, H, K (see (6.2.3) and (6.2.15)). Finally, since both two-scale matrices $P = P_m$ and $Q = Q_m$ can be easily computed, we can also find the decomposition matrices $A = A_m$ and $B = B_m$ by applying (6.2.18) in Theorem 6.2, as well as the dual B-splines and dual B-wavelets by applying (6.2.32) for $\{\widetilde{u}_i\}$ and either (6.2.32) or (6.2.30) for $\{\widetilde{v}_i\}$ and $\{\widetilde{w}_i\}$.

Example 6.15. Let $f(t) = \frac{(t-4)|t-4|}{4} + 4$, and consider the knot sequences $\mathbf{x}$ and $\mathbf{t}$ in (6.7.1) on the interval [0,8] with $n = 8$ and

$$\begin{cases} x_i = f(i), \text{ for } i = 0, \ldots, 8, \text{ and} \\ x_{-3} = x_{-2} = x_{-1} = 0, \quad x_9 = x_{10} = x_{11} = 8; \end{cases}$$

$$\begin{cases} t_i = f\left(\dfrac{i}{2}\right), \text{ for } i = 0, \ldots, 16, \text{ and} \\ t_{-3} = t_{-2} = t_{-1} = 0, \quad t_{17} = t_{18} = t_{19} = 8. \end{cases}$$

Hence, there are eight wavelets

$$w_i = \psi_{\mathbf{x},\mathbf{t};m,i-m}, \quad i = 1, \ldots, 8,$$

independent of the order m. For $m = 2$, there are two boundary wavelets, w_1 and w_8, and six interior ones. Their graphs are shown in Figures 6.7–6.14. For $m = 4$, there are six boundary wavelets, w_1, w_2, w_3 and w_6, w_7, w_8, and two interior ones, w_4, w_5. Their graphs are shown in Figures 6.15–6.22.

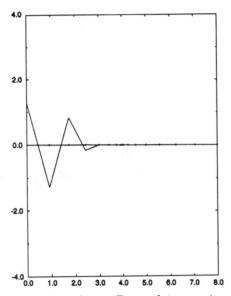

FIG. 6.7.   *Boundary linear B-wavelet* $w_1 = \psi_{\mathbf{x},\mathbf{t};2,-1}(t)$.

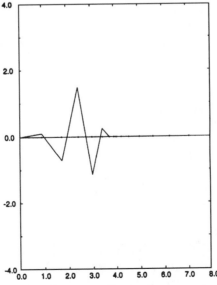

FIG. 6.8.   *Interior linear B-wavelet* $w_2 = \psi_{\mathbf{x},\mathbf{t};2,0}(t)$.

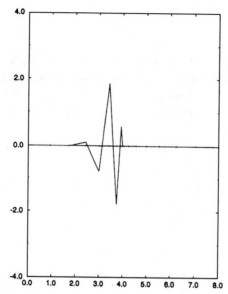

FIG. 6.9.   *Interior linear B-wavelet* $w_3 = \psi_{\mathbf{x},\mathbf{t};2,1}(t)$.

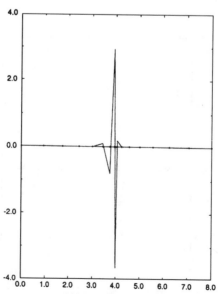

FIG. 6.10.   *Interior linear B-wavelet* $w_4 = \psi_{\mathbf{x},\mathbf{t};2,2}(t)$.

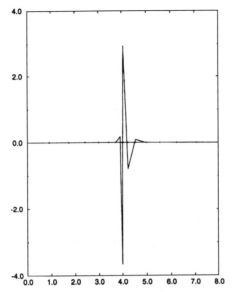

FIG. 6.11. *Interior linear B-wavelet $w_5 = \psi_{\mathbf{x},\mathbf{t};2,3}(t)$.*

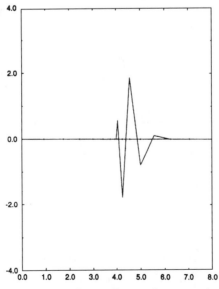

FIG. 6.12. *Interior linear B-wavelet $w_6 = \psi_{\mathbf{x},\mathbf{t};2,4}(t)$.*

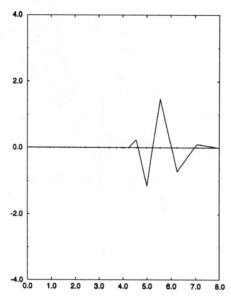

FIG. 6.13.   *Interior linear B-wavelet $w_7 = \psi_{\mathbf{x},\mathbf{t};2,5}(t)$.*

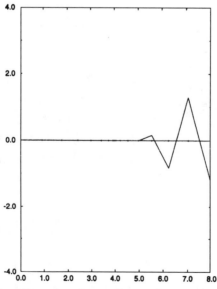

FIG. 6.14.   *Boundary linear B-wavelet $w_8 = \psi_{\mathbf{x},\mathbf{t};2,6}(t)$.*

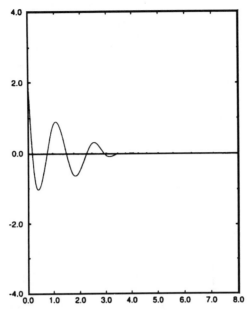

FIG. 6.15. *Boundary cubic B-wavelet* $w_1 = \psi_{\mathbf{x},\mathbf{t};4,-3}(t)$.

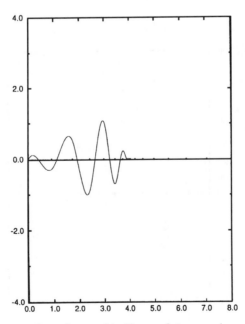

FIG. 6.16. *Boundary cubic B-wavelet* $w_2 = \psi_{\mathbf{x},\mathbf{t};4,-2}(t)$.

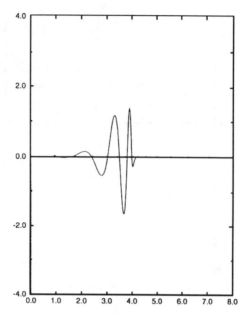

FIG. 6.17.   *Boundary cubic B-wavelet $w_3 = \psi_{\mathbf{x},\mathbf{t};4,-1}(t)$.*

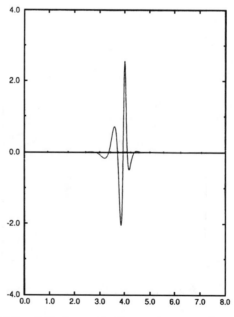

FIG. 6.18.   *Interior cubic B-wavelet $w_4 = \psi_{\mathbf{x},\mathbf{t};4,0}(t)$.*

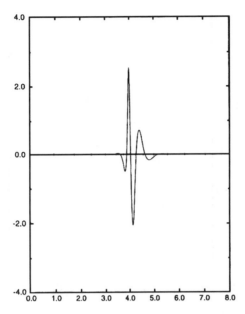

FIG. 6.19. *Interior cubic B-wavelet* $w_5 = \psi_{\mathbf{x},\mathbf{t};4,1}(t)$.

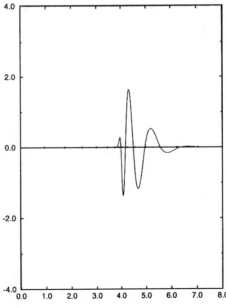

FIG. 6.20. *Boundary cubic B-wavelet* $w_6 = \psi_{\mathbf{x},\mathbf{t};4,2}(t)$.

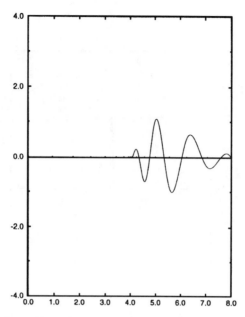

FIG. 6.21.   *Boundary cubic B-wavelet* $w_7 = \psi_{\mathbf{x},\mathbf{t};4,3}(t)$.

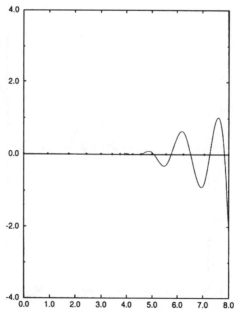

FIG. 6.22.   *Boundary cubic B-wavelet* $w_8 = \psi_{\mathbf{x},\mathbf{t};4,4}(t)$.

CHAPTER 7

Applications

The wavelet transform has the capability of bringing out certain special features of a function under investigation. For instance, isolated discontinuities of the mth-order derivative of a function can be easily detected by the integral wavelet transform by using an analyzing wavelet with $m+2$ vanishing moments. If a function represents an acoustic signal, the wavelet transform separates the signal into different octaves and identifies both time locations and magnitudes of the acoustic notes in each frequency octave band. If the function represents a two-dimensional image, then the wavelet transform brings out the spatial and spectral redundancies of the image, again in each octave band but in each of the horizontal and vertical directions. When a differential or integral equation is considered, the function under investigation is the solution of the equation. The wavelet transform can then be used to identify the components of the solution with least significance, so that ignoring these components greatly accelerates numerical computation of the solution. This chapter is intended to highlight these aspects of the wavelet transform.

7.1. Detection of singularities and feature extraction.

It has been pointed out in section 6.5 of Chapter 6 that if a function $f(t)$ has a Taylor expansion at $t = t_0$ as given by (6.5.1), then its integral wavelet transform (IWT) with analyzing wavelet $\psi(t)$ at t_0 with scale $a > 0$ is the same as the IWT of the remainder $R_m(t)$ of this Taylor expansion, provided that $\psi(t)$ has m vanishing moments and that the support (i.e., time duration) of $\psi(\frac{t-t_0}{a})$ is a subset of the interval $(t_0 - \delta, t_0 + \delta)$. Since $R_m(t)$ is usually relatively smaller than $f(t)$ in this interval, we see that the value of

$$(W_\psi f)(t_0, a) \qquad (7.1.1)$$

is usually negligible. Now let $\delta > 0$ be the radius of the Taylor expansion at t_0. That is, the mth-order derivative $f^{(m)}(t)$ of $f(t)$ has a discontinuity at $t_0 \pm \delta$. Then we see that the value of the IWT

$$(W_\psi f)(t_1, a) \qquad (7.1.2)$$

161

is significantly larger than the value in (7.1.1) if t_1 is close to $t_0 - \delta$ or $t_0 + \delta$. In other words, the IWT is capable of "detecting" the "singularity" of $f(t)$ at $t_0 - \delta$ or $t_0 + \delta$. Obviously the detection is more accurate if the support of $\psi(t)$ is smaller. Here, "singularities" mean isolated discontinuities of any of the derivatives $f^{(k)}(t), 0 \le k \le m$.

7.1.1. Detection of isolated singularities.

In Chapter 1, we demonstrated the zoom-in/zoom-out property of the IWT using the example of some music data with additive pop noise in Figure 1.15. Observe that the locations of the occurrence of the pop noise show up very clearly while the Fourier transform is not capable of detecting the pop noise at all. It is interesting (and important) to point out that the pop noise can be viewed even more clearly in the octave bands as represented by the individual wavelet series with coefficients given by the discrete wavelet transform (DWT) as shown in Figure 1.17. This is perhaps not very surprising since the DWT is simply the set of IWT values at the dyadic locations $t = \frac{k}{2^j}$, and other values of the IWT (hopefully with smaller magnitudes) are not shown at all. Note that we are careful in saying here that hopefully only the other IWT values are not shown in the DWT. This may not be the case, and we will return to a discussion of this type of behavior of the DWT after we give two more examples.

Let us return again to Chapter 1 and try to find the locations of the small perturbations on the sinusoidal curve in Figure 1.19. This graph is actually a linear spline representation of a perturbed sinusoidal with $2^9 - 1$ equally spaced knots on each unit interval (i.e., the linear spline representation in the multiresolution analysis (MRA) space V_j with $j = 9$). The first level wavelet decomposition of the curve (i.e., the wavelet component in W_8) is shown in Figure 1.20. Observe that this octave band with DWT coefficients provides a very accurate detection of the perturbations.

To give a more dramatic example, we consider a curve consisting of two cubic polynomials joined together in a C^1 fashion as shown in the top right corner of Figure 7.1. That is, this curve consists of two cubic polynomials and has continuous turning tangents, but the second derivative has a jump discontinuity somewhere. In other words, this curve does not have a cardinal cubic spline series representation. The cardinal cubic spline interpolation of this curve from V_8 (i.e., $2^8 - 1$ knots on each unit interval) is shown at the top left corner in Figure 7.1. Recall that two interpolation schemes were given in Examples 6.3–6.4 in Chapter 6. Now, the second through fourth rows of Figure 7.1 show the first, second, and third levels of wavelet decomposition, where the wavelet components have been magnified. Observe that the location of the discontinuity of $f''(t)$ shows up very clearly.

7.1.2. Shift variance of the DWT.

We remarked in the beginning of the previous subsection that in applying the DWT, there is the danger of missing certain values of the IWT that are not located at the dyadic positions. To demonstrate this so-called shift-variance behavior of the DWT, let us consider the following Haar example.

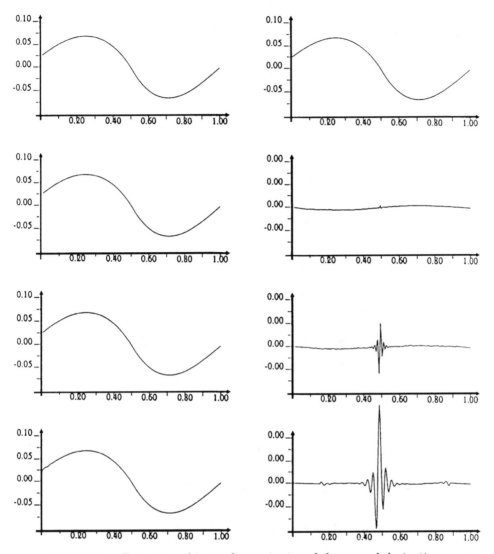

FIG. 7.1. *Detection of jump discontinuity of the second derivative.*

Example 7.1. Recall that the Haar decomposition sequences are given by $\{\frac{1}{2}, \frac{1}{2}\}$ and $\{\frac{1}{2}, -\frac{1}{2}\}$. Hence, to detect the change of the data sequence

$$c_{0,\ell} = \delta_{\ell,1} + \delta_{\ell,2}$$

at $\ell = 1$ and 2, where $\delta_{\ell,k}$ denotes the Kronecker delta, we apply the wavelet decomposition sequence $b_0 = \frac{1}{2}, b_1 = -\frac{1}{2}$ (and $b_k = 0$ otherwise) to obtain the DWT

$$d_{-1,k} = \sum_{\ell} b_{\ell-2k} c_{0,\ell} = \sum_{\ell} b_{\ell-2k} \delta_{\ell,1} + \sum_{\ell} b_{\ell-2k} \delta_{\ell,2}$$

$$= b_{1-2k} + b_{2-2k} = \frac{1}{2}(\delta_{k,1} - \delta_{k,0}).$$

Observe that the detection misses the target position by only one time unit. However, if the data sequence $\{c_{0,\ell}\}$ is shifted to the left by one time unit, becoming

$$\tilde{c}_{0,\ell} = \delta_{\ell,0} + \delta_{\ell,1},$$

then the DWT becomes

$$\tilde{d}_{-1,k} = \sum_{\ell} b_{\ell-2k}\tilde{c}_{0,\ell} = \sum_{\ell} b_{\ell-2k}\delta_{\ell,0} + \sum_{\ell} b_{\ell-2k}\delta_{\ell,1}$$

$$= b_{-2k} + b_{1-2k} = \frac{1}{2}(\delta_{k,0} - \delta_{k,0}) = 0$$

for all k. So, the one-level DWT fails to detect the change at $\ell = 0$ and 1 of the shifted sequence $\{\tilde{c}_{0,\ell}\}$.

This phenomenon of not necessarily matching the shift of the one-level DWT with the one-level DWT of the same shift of a data sequence is called *shift variance* (or translation variance) of the (discrete) wavelet decomposition.

To avoid shift variance, a simple method is perhaps to skip the downsampling operation. This works well, at least for the first level DWT, as follows. Let $\phi(t)$ and $\psi(t)$ be a scaling function and wavelet pair with corresponding duals $\tilde{\phi}(t)$ and $\tilde{\psi}(t)$. Consider the one-unit shift

$$f_n^s(t) = f_n\left(t - \frac{1}{2^n}\right)$$

of a function $f_n(t) \in V_n$. It is clear that $f_n^s(t)$ is in V_n also. Hence, we may write

$$\begin{cases} f_n(t) = \sum_k c_{n,k}\phi(2^n t - k), \\ f_n^s(t) = \sum_k c_{n,k}^s \phi(2^n t - k), \end{cases}$$

and the dual property of $\phi(t)$ and $\tilde{\phi}(t)$ gives

$$c_{n,k}^s = 2^n \int_{-\infty}^{\infty} f^s(t)\overline{\tilde{\phi}(2^n t - k)}dt$$

$$= 2^n \int_{-\infty}^{\infty} f\left(t - \frac{1}{2^n}\right)\overline{\tilde{\phi}(2^n t - k)}dt \qquad (7.1.3)$$

$$= 2^n \int_{-\infty}^{\infty} f(t)\overline{\tilde{\phi}(2^n t - k + 1)}dt$$

$$= c_{n,k-1}.$$

Recall the wavelet decomposition algorithm

$$d_{n-1,k} = \sum_{\ell} b_{\ell-2k}c_{n,\ell} \qquad (7.1.4)$$

from (5.3.24). So, without downsampling, we have

$$D_{n-1,k} = \sum_{\ell} b_{\ell-k} c_{n,\ell},$$
                                                                    (7.1.5)

so that

$$D_{n-1,2m} = d_{n-1,m}.$$
                                                                    (7.1.6)

That is, for even k, say $k = 2m$, we return to the DWT with downsampling. Now for odd k, say $k = 2m - 1$, we have, from (7.1.5), (7.1.3), and (7.1.4), consecutively,

$$D_{n-1,2m-1} = \sum_{\ell} b_{\ell-2m+1} c_{n,\ell} = \sum_{\ell} b_{\ell-2m} c_{n,\ell-1}$$

$$= \sum_{\ell} b_{\ell-2m} c_{n,\ell}^s = d_{n-1,m}^s.$$

That is, for odd values of k, say $k = 2m - 1$, we have

$$D_{n-1,2m-1} = 2^{(n-1)/2} (W_{\widetilde{\psi}} f^s) \left(\frac{m}{2^{n-1}}, \frac{1}{2^{n-1}} \right)$$

$$= 2^{(n-1)/2} (W_{\widetilde{\psi}} f) \left(\frac{m - \frac{1}{2}}{2^{n-1}}, \frac{1}{2^{n-1}} \right).$$

Combining (7.1.6) and (7.1.7), we see that without downsampling, the decomposition procedure (7.1.5) yields

$$D_{n-1,k} = 2^{(n-1)/2} (W_{\widetilde{\psi}} f) \left(\frac{k}{2^n}, \frac{1}{2^{n-1}} \right).$$
                                                                    (7.1.7)

These are values of $2^{(n-1)/2}$ multiple of the IWT of $f(t)$ at the scale $a = \frac{1}{2^{n-1}}$, but at the same locations $b = \frac{k}{2^n}$ as the sample points before the decomposition (7.1.5) is applied.

7.1.3. Choice of analyzing wavelets for detection.

Another very important point to consider in providing an effective detection tool is the choice of analyzing wavelets. From our discussion at the beginning of this section, we recall that a desirable analyzing wavelet should have small support (i.e., time duration) and as many vanishing moments as possible. These two requirements are competing, and so there must be a compromise. From the time-frequency localization point of view, the optimal criterion is the uncertainty lower bound $\frac{1}{2}$, which as we recall from (2.4.20) and (5.4.9) is sharp but cannot be achieved by any wavelet. In fact, if the mth-order cowlets $\Psi_m(t)$ in (5.4.8) are used as analyzing wavelets, then although the size of time-frequency windows decreases to the uncertainty lower bound $4 \times \frac{1}{2} = 2$, the supports $[0, n_m]$ or filter lengths $n_m + 1$ (with $n_m \geq m$) increase as m tends to infinity. (See section 5.4 and particularly Theorem 5.13 in Chapter 5. Recall also that $\Psi_m(t)$ has m vanishing moments.) The required number of vanishing moments depends on the detection problem. For targets that are somewhat obvious (such as a jump discontinuity) even the linear spline wavelet with two vanishing moments is sufficient; but for the less obvious ones (such as cracks in Figure 7.1), the number of vanishing moments must be at least 4.

7.1.4. Spline-wavelet detection procedure.

Suppose that a semiorthogonal wavelet such as a cowlet $\Psi_m(t)$ is used as the analyzing wavelet for a detection problem. Then the length of the bandpass filter $\{b_k\}$ for wavelet decomposition is $n_m + 1$, and the shortest one with $n_m = m$ is achieved by the mth-order B-wavelet $\psi_m(t)$. B-wavelets of order two (for linear spline wavelet $\psi_2(t)$) through order five (for quintic spline wavelet $\psi_6(t)$) are shown in Figures 5.2–5.6. In order to use such wavelets as analyzing wavelets, the function (or signal) to be analyzed must be represented as a *dual* B-spline series, with scaling function $\widetilde{N}_m(t)$ as shown in Figures 3.9–3.11 in Chapter 3.

The first step is to represent the function (or signal) as a B-spline series with a sufficiently fine knot sequence (i.e., sufficiently large bandwidth). Methods such as *orthogonal projection* (discussed in section 3.1, (3.1.20)–(3.1.23) in Chapter 3), *interpolation* (discussed in section 6.1, (6.1.3)–(6.1.9) in Chapter 6), and *quasi interpolation* (see Example 6.5, (6.1.16)–(6.1.16) in Chapter 6) are popular choices. So, for the MRA nested sequence $\{V_n\}$ of spline spaces generated by the mth-order B-spline $N_m(t)$, we have the B-spline series representation

$$f_n(t) = \sum_k c_{n,k} N_m(2^n t - k) \tag{7.1.8}$$

of the original signal $f(t)$.

The second step is to change the B-spline series representation (7.1.8) to the dual B-spline series representation

$$f_n(t) = \sum_k \widetilde{c}_{n,k} \widetilde{N}_m(2^n t - k). \tag{7.1.9}$$

Recall from (3.4.21) that $\{N_m(t-k)\}$ and $\{\widetilde{N}_m(t-k)\}$ are biorthogonal families. Hence, since

$$\langle N_m(2^n t - k), N_m(2^n t - \ell) \rangle = 2^{-n} N_{2m}(m + \ell - k) \tag{7.1.10}$$

(see (4.3.1) in Chapter 4), we have the simple formula

$$\widetilde{c}_{n,k} = \sum_\ell N_{2m}(m + k - \ell) c_{n,\ell} \tag{7.1.11}$$

for computing the coefficient sequence of the dual B-spline series representation (7.1.9) (see (5.3.35)).

Example 7.2. For dual linear B-spline series representations (i.e., $m = 2$), we have

$$\tilde{c}_{n,k} = \frac{1}{6}(c_{n,k-1} + 4c_{n,k} + c_{n,k+1}). \tag{7.1.12}$$

For dual quadratic B-spline series representations (i.e., $m = 3$), we have

$$\tilde{c}_{n,k} = \frac{1}{120}(c_{n,k-2} + 26c_{n,k-1} + 66c_{n,k} + 26c_{n,k+1} + c_{n,k+2}). \tag{7.1.13}$$

For dual cubic B-spline series representations (i.e., $m = 4$), we have

$$\tilde{c}_{n,k} = \frac{1}{5040}(c_{n,k-3} + 120c_{n,k-2} + 1191c_{n,k-1} + 2416c_{n,k} \\ + 1191c_{n,k+1} + 120c_{n,k+2} + c_{n,k+3}). \tag{7.1.14}$$

For dual quartic B-spline series representations (i.e., $m = 5$), we have

$$\tilde{c}_{n,k} = \frac{1}{362880}(c_{n,k-4} + 502c_{n,k-3} + 14608c_{n,k-2} + 88234c_{n,k-1} \\ + 156190c_{n,k} + 88234c_{n,k+1} + 14608c_{n,k+2} + 502c_{n,k+3} + c_{n,k+4}). \tag{7.1.15}$$

The third step is to apply the wavelet decomposition algorithm (5.3.29) with filter sequences $\{p_k\}$ and $\{q_k\}$ given by (5.3.36), where the values of $N_{2m}(m + k - \ell + 1)$ for $\{w_k\}$ in (5.3.36) are precisely the coefficients shown in (7.1.12)–(7.1.15) for $m = 2, \ldots, 5$, respectively. That is, we have

$$\begin{cases} \tilde{c}_{j-1,k} = \sum_{\ell} p_{2k-\ell}\tilde{c}_{j,\ell}, \\ \tilde{d}_{j-1,k} = \sum_{\ell} q_{2k-\ell}\tilde{c}_{j,\ell} \end{cases} \tag{7.1.16}$$

for $j = n, n - 1, \ldots$.

The final result is (certain constant multiple of) the DWT of the spline representation $f_n(t)$ in (7.1.8) of the original signal $f(t)$, with the B-spline wavelet $\psi_m(t)$ as the analyzing wavelet. More precisely, we have

$$\left(W_{\psi_m} f_n\right)\left(\frac{k}{2^j}, \frac{1}{2^j}\right) = 2^{-j/2}\tilde{d}_{j,k}. \tag{7.1.17}$$

If we wish to compute the wavelet transform of $f_n(t)$ at the original sample points $\frac{k}{2^n}$, $k = 0, \pm 1, \ldots$, then the downsampling operation is ignored. The first-level decomposition was discussed in the subsection 7.1.2. However, one has to proceed with care in the lowpass decomposition in computing the IWT values at $\frac{k}{2^n}$ for the lower (scale) levels.

7.1.5. Two-dimensional feature extraction.

When a function of two variables is studied, detection of "singularities" (or irregularities) of this function from any direction is possible. The multidimensional wavelet transform was introduced in section 6.4 of Chapter 6, where a tensor-product wavelet $\psi(x_1) \cdots \psi(x_s)$ was used as the analyzing wavelet. An advantage of this type of wavelet transform is that computation of the transform reduces to repeated computations of one-dimensional wavelet transforms. For instance, for one-level decomposition, the two-dimensional DWT was separated into four filtering equations described in Algorithm 6.3 in section 3 of Chapter 6, and these four filters can be realized in two stages, first in the x-direction and then in the y-direction as shown in Figure 7.2, yielding the LL, LH, HL, and HH components as described in Figure 6.2.

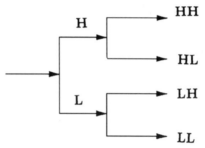

FIG. 7.2. *Implementation of the two-dimensional decomposition algorithm.*

Recall that the LL band of the scaling function (e.g., B-spline) series representation for some MRA subspace V_n of the original function is the orthogonal projection of this representation function to the immediate subspace V_{n-1} of the MRA. Hence, if the function describes an image, then this LL band is the same image but with $\frac{1}{4}$ of the original resolution. More precisely, the high-frequency details of the image are revealed in the other three bands, with "corners" revealed in the HH band, horizontal "edges" in the LH band, and vertical "edges" in the HL band. Of course, some corners may be missed and most edges are not complete. This is why we need more than one level of decomposition. In fact, the details of corners and edges are usually not visually obvious, unless the small values of the DWT are removed (this process is called "thresholding") and the details (i.e., larger values of the DWT) are magnified. In Figure 7.3, we show the original of the "house" image on the left and the first level wavelet decomposition on the right. The pixel values in the HL, LH, and HH bands in Figure 7.3 are first thresholded and then magnified, and the much more obvious details are shown in Figure 7.4. Next the image size of the LL band is magnified four times using pixel replication and is treated as an original image; the same process is repeated. The results are shown in Figures 7.5 and 7.6. Finally, all the thresholded and magnified details in Figure 7.4 and Figure 7.6 are put together and the result is shown in Figure 7.7. Observe that extraction of the horizontal and vertical features shown in Figure 7.7 of the house image is quite satisfactory, but the other features, such as the diagonal edges, do not show up very well.

FIG. 7.3. *The house image and its first-level wavelet decomposition.*

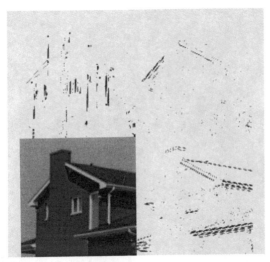

FIG. 7.4. *Thresholded and magnified details of the first-level DWT.*

FIG. 7.5. *The LL subimage is magnified four times using pixel replication and is treated as an original image, and the DWT of this subimage is shown on the right.*

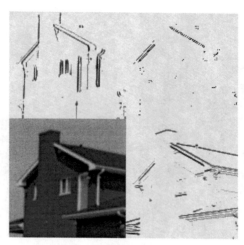

FIG. 7.6.   *Thresholded and magnified details of the second-level DWT.*

FIG. 7.7.   *Superposition of the details from Figures 7.4 and 7.6.*

In order to better reveal the edge features, perhaps the first approach that comes to mind is to rotate the picture before applying the two-dimensional feature extraction procedure discussed above. However, there are some diffi-culties associated with this approach. The most serious one is that the two-dimensional wavelet decomposition algorithm only applies to pixel values on a rectangular grid. To overcome this difficulty, we could compute new pixel values by evaluation of the scaling function (e.g., B-spline) series representa-tion on the new rectangular grid. If a cardinal B-spline is used as the scaling function, then the spline evaluation algorithm described in Theorem 3.3(viii) in Chapter 3 can be used. Otherwise, the usual table look-up procedure can be applied. It should be noted that the boundaries of the picture are no longer horizontal and vertical, so a more extensive extension across the boundaries is required. Extension by reflection is usually most effective. Fortunately for feature extraction, only a portion of the image is of interest, and focusing on a rectangular subimage may be sufficient already.

7.2. Data compression.

We have discussed in the previous section the capability of the wavelet transform for detecting changes and extracting image features. In the "house" example in subsection 7.1.5, for instance, the details of the image contain a large amount of small DWT values. These are the high-frequency pixel values that were discarded in the thresholding process. They do not contribute much to the visual quality of the original image. In other words, by putting the "corners" and "edges" back to the LL band, we have almost recovered the original image. Hence, if the corner and edge details occupy a very small percentage (usually less than 5%) of the image file, we have already obtained almost 4 to 1 and 16 to 1 compression ratios for one-level and two-level decompositions, respectively. So, if other techniques, particularly "entropy coding" (to be discussed in this section) are included, higher compression ratios can be achieved. This section is devoted to a discussion of data compression by using the wavelet transform along with other compression techniques.

7.2.1. The compression scheme.

In order to apply the wavelet transform or, more precisely, the DWT to data compression, the first step is to map the original signal (or image) to be compressed to an MRA subspace, say V_n, with appropriately large n. This procedure was outlined in section 6.1 of Chapter 6. For the one-dimensional setting, we now have the coefficient sequence $\{c_{n,k}\}$, as in (6.1.1), and for the two-dimensional setting, $\{c_{LL,k_1,k_2}^n\}$, as in (6.4.16). For convenience, we will use the notation $\mathbf{c}_n$ to represent both of them. In addition, we will use the notation f to represent $f(t)$ for a one-dimensional signal and $f(x,y)$ for a two-dimensional one such as an image. Hence, the first step is to map the original signal f, whether in digital or analog format, to a digital representation $\mathbf{c}_n$; namely,

$$f \longrightarrow \mathbf{c}_n. \tag{7.2.1}$$

Suppose that the scaling function that facilitates the mapping procedure (7.2.1) is $\phi(t)$ or $\phi(x)\phi(y)$, and that the corresponding wavelet $\psi(t)$ is the desirable analyzing wavelet. Then $\mathbf{c}_n$ is the coefficient sequence of the dual scaling function (or dual B-spline) series representation in terms of $\widetilde{\phi}(t)$ or $\widetilde{\phi}(x)\widetilde{\phi}(y)$, instead of the "modeling" scaling function. In other words, an extra step of changing the series representation from ϕ to $\widetilde{\phi}$ as discussed in (7.1.11) of subsection 7.1.4 is required to accomplish the transformation (7.2.1). Of course, this effort is not necessary if $\phi(t)$ is an orthonormal scaling function, and is not possible if $\psi(t)$ is not a semiorthogonal wavelet. (See the discussion between (5.1.18) and (5.1.19) in Chapter 5.) If the dual wavelet $\widetilde{\psi}(t)$ is the desirable analyzing wavelet, then the extra step mentioned above is not required. For convenience, let us assume in the following that the dual wavelet $\widetilde{\psi}(t)$ (or $\widetilde{\phi}(x)\widetilde{\psi}(y), \widetilde{\psi}(x)\widetilde{\phi}(y),$ $\widetilde{\psi}(x)\widetilde{\psi}(y)$ in the two-dimensional setting) is chosen as the analyzing wavelet.

The second step is to apply the wavelet decomposition algorithm for any desirable number of levels. (See (1.3.11), (5.3.9), (6.2.21), and in the two-dimensional setting, Algorithm 6.3 and its implementation in Figure 7.2.) Regardless of the number of levels of decompositions and whether one- or two-dimensional settings are considered, we will use the schematic diagram shown in Figure 7.8, where $m < n$, $n - m$ is the number of levels of wavelet decompositions, and $\mathbf{d}_{m,n}$ denotes the DWT obtained as a result of the decompositions.

FIG. 7.8.   *Wavelet transform with analyzing wavelet $\widetilde{\psi}$.*

The third step of this compression scheme is *quantization*. This is the first (and in most applications, the only) lossy (i.e., nonreversible) transformation of the entire wavelet compression procedure. Depending on the choice of the analyzing wavelet in the wavelet transform, quantization could greatly reduce the "redundant" information of the output sequences $\mathbf{c}_m$ and $\mathbf{d}_{m,n}$ of the wavelet decomposition (shown in Figure 7.8). This is achieved by a "many-to-one" map in the sense that many values of the sequences $\mathbf{c}_m$ and $\mathbf{d}_{m,n}$ are replaced by one (quantized) integer value. In particular, many small values of the DWT $\mathbf{d}_{m,n}$ are replaced by zero. So, if the analyzing wavelet has small support (or time duration) and a large number of vanishing moments, the number of nonzero values of the quantized DWT $d_{m,n}$ is very small. In other words, quantization also gives rise to thresholding used in singularity detection and feature extraction. The effectiveness of the analyzing wavelet for the purpose of quantization is reflected in the shapes of the "histograms" of each of the octave bands. For typical test signals (or images), these histograms should have the shape of a Laplacian curve, centered at the origin and having small standard deviation (see Figure 7.9 for a typical example).

The many-to-one map that describes a quantizer is a step (i.e., staircase) function with integer range $\{y_i\}$ as shown in Figure 7.10. Note that the step $y = y_0 = 0$ is usually centered at the origin. Hence, the thresholding parameter is half of the size of this ground level step of the quantizer.

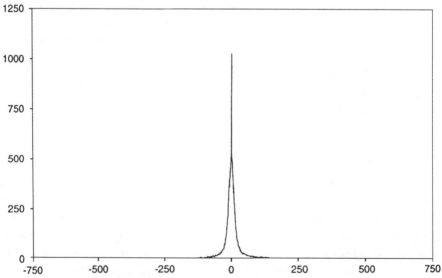

FIG. 7.9.   *Histogram of an octave band of the DWT of a typical signal.*

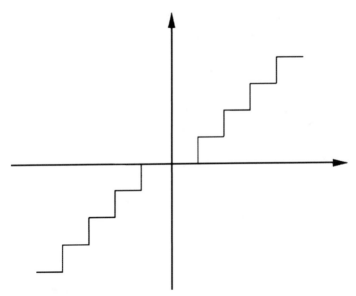

FIG. 7.10.   *A many-to-one quantization map.*

There are several quantization schemes. The most popular ones are scalar quantization (in which each value is individually quantized), vector quantization (that quantizes several values together), and predictive quantization (which quantizes the difference between a sample value and a prediction of this value as well). For wavelet compression, scalar quantization is probably the best because of its simplicity. However, the optimal scalar quantizer, where optimality is characterized by least-squared distance from the quantized integer values y_i on each step (or decision) interval $[x_i, x_{i+1})$ using the Laplacian distribution to define the distance metric, is also computationally expensive. For this reason, *uniform quantization*, in which

$$x_{i+1} - x_i = h > 0 \tag{7.2.2}$$

for all i, is often used. The step size h of a uniform quantizer (7.2.2) is also the thresholding parameter.

When scalar quantization is used, whether it is optimal, uniform, or nonuniform, the integers y_i, with

$$y_{-m} < \cdots < y_{-1} < y_0 = 0 < y_1 < \cdots < y_n, \tag{7.2.3}$$

that constitute the range of a scalar quantizer are to be *coded* to gain a substantial improvement of the overall compression. This fourth and final step of the wavelet compression scheme is called *entropy coding*. It is a lossless (i.e., reversible) compression procedure. We remark without going into any details, however, that, for two-dimensional signals such as images, the quantized integer values y_i can first be arranged as a certain "tree" before an entropy encoding procedure is performed. Since the branches of the tree are sometimes modified for efficiency purposes, this compression step, to a certain degree, could be designed to be lossy. There is the so-called embedded zero-tree wavelet (EZW) of Shapiro or a more recent improvement by Said and Pearlman.

Suppose that the totality of all values (consisting of the DWT values in $\mathbf{d}_{m,n}$ and the DC terms in $\mathbf{c}_m$) to be quantized is denoted by

$$\{c_j\}, \quad j \in J. \tag{7.2.4}$$

In general, this is a very large set of redundant real numbers. The quantization procedure maps this set to a much smaller set

$$\{y_i\}, \quad i \in I \tag{7.2.5}$$

of integers, which is distinct and, in fact, forms an increasing sequence given by (7.2.3). Since $\{y_i\}$ is to be coded, we call

$$I = \{i\} \leftrightarrow \{y_i\} \tag{7.2.6}$$

a codebook and each $i \leftrightarrow y_i$ a symbol. However, since quite a few of the values in $\{c_j\}$ are mapped to each y_i by the quantizer in general, the *codewords* assigned to the symbols in the codebook must allow the user to decode any message in a unique way, when some of the codewords are repeated and mixed together to form the message. For instance, suppose the codebook consists of four symbols, $y_0, \ldots, y_3$, and the following "codewords" are assigned to them:

$$
\begin{aligned}
y_0 &\text{ ——— } 0, \\
y_1 &\text{ ——— } 1, \\
y_2 &\text{ ——— } 10, \\
y_3 &\text{ ——— } 100.
\end{aligned}
\tag{7.2.7}
$$

Also, suppose that five real numbers, $x_0, \ldots, x_4$, were quantized with two values mapped to y_0 and the remaining mapped to the remaining symbols y_1, y_2, y_3. Now the message

$$10011000 \tag{7.2.8}$$

cannot be decoded (uniquely), since it may mean

$$y_2 y_0 y_1 y_3 y_0$$

or

$$y_3 y_1 y_2 y_0 y_0,$$

etc. On the other hand, the codewords

$$
\begin{aligned}
y_0 &\text{ ——— } 1, \\
y_1 &\text{ ——— } 01, \\
y_2 &\text{ ——— } 001, \\
y_3 &\text{ ——— } 000
\end{aligned}
\tag{7.2.9}
$$

are truly codewords, since they allow (unique) decoding of any message formed by an arbitrary arrangement of repeated codewords. For instance, for the codewords in (7.2.9), the message in (7.2.8) represents precisely

$$y_0 y_2 y_0 y_3$$

without any ambiguity.

Entropy coding is used to assign codewords to the symbols to form a codebook according to the frequency of appearance of the values c_j in (7.2.4) mapped to the symbols by the quantizer. More precisely, let

$$q_i = \text{ number of } c_j, \quad j \in J,$$
$$\text{mapped to } y_i \text{ by the quantizer.} \tag{7.2.10}$$

This number can be easily obtained from the histogram. We normalize the q_i's by introducing

$$p_i = \frac{q_i}{\sum_{l \in I} q_l}. \tag{7.2.11}$$

Hence, $0 \le p_i \le 1$, and for each $i \in I$, p_i may be considered as the probability of occurrence of the symbol y_i in any code.

The most popular entropy coding schemes are Huffman and arithmetic codings. Both codings are statistical schemes. The Huffman coding is optimal as governed by the theoretical entropy lower bound if (and only if) the probabilities p_i are integer powers of $\frac{1}{2}$. Since this rarely happens, arithmetic coding usually gives better compression results but at the expense of a much larger amount of computing resources both in terms of CPU time and memory.

The Huffman coding is fairly simple. It is constructed by building a binary tree for the symbols y_i according to their probabilities p_i of occurrence in a message. Let us demonstrate this coding scheme again by considering a codebook with four symbols $y_0, \ldots, y_3$ with probabilities $p_0, \ldots, p_3$, respectively. The first thing to do is to pick the least probable two, say y_2 and y_3, and form a composite symbol which is a two-symbol message $y_3 y_2$. Assign $p_2 + p_3$ as the probability of this composite symbol, and assign 0 to the most probable half and 1 to the other half of the composite symbol $y_3 y_2$. Let us assume that y_3 is the most probable one, which is then labeled 0. This process is repeated with y_2 and y_3 replaced by the composite symbol $y_3 y_2$ with combined probability. This reduces the size of the codebook by 1. Suppose that $p_2 + p_3$ satisfies $p_1 \le p_2 + p_3 \le p_0$. Then y_1 and $y_3 y_2$ constitute the next pair, and the composite symbol is labeled 0 while y_1 is labeled 1. Now the final step is to compare the probability p_0 of y_0 and the combined probability $p_1 + p_2 + p_3$ of the new composite symbol (or three-symbol message) $y_3 y_2 y_1$. If $p_1 + p_2 + p_3 \ge p_0$, then we have the binary tree shown in Figure 7.11. The codewords can be read off by following the branches of the binary tree. Hence, we obtain the codes in (7.2.9).

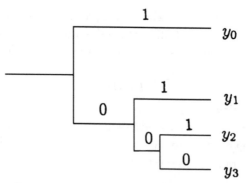

FIG. 7.11.   *A Huffman binary tree.*

As another example, if $p_2 \leq p_3 \leq p_0 \leq p_1$, but $p_1 \leq p_2 + p_3$, then the binary tree for the Huffman coding is shown in Figure 7.12, and the codewords for the codebook are given by (7.2.12).

$$
\begin{aligned}
y_0 &\text{——} 01, \\
y_1 &\text{——} 00, \\
y_2 &\text{——} 11, \\
y_3 &\text{——} 10.
\end{aligned}
\qquad (7.2.12)
$$

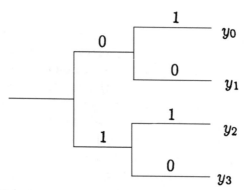

FIG. 7.12.   *Another Huffman binary tree.*

Observe that the Huffman binary tree is not unique since several symbols or composite symbols may have the same probabilities. However, all Huffman binary trees for a codebook (with given probabilities) have the same compressed length.

The arithmetic coding, on the other hand, is constructed by partitioning the probability interval $[0, 1]$ according to the probabilities of the symbols in such a way that the lengths of the subintervals are given by the probability values and that they are arranged in order of decreasing lengths. The subintervals are again partitioned according to the ratios (hence, products) of the probabilities. This yields two-symbol messages. As a message gets longer, the interval representing this message gets smaller, and the number of bits needed to specify the interval increases. Hence, the coding scheme terminates according to the bit limitation.

This completes our discussion of wavelet data compression. The decompression procedure is simply the reverse of the compression procedure described above. The only difference is that since the quantization step function does not have an inverse, certain signal quality is lost. However, if a good analyzing wavelet is used so that the histograms of the DWT are very narrow Laplacian curves, then most of the DWT values were already replaced with zero by the quantizer and there is no need to reverse these thresholded values. Consequently, very few nonzero values require "dequantization."

7.2.2. Audio compression.

Wavelet data compression is achieved in two stages, namely, quantization and entropy coding. The first stage is lossy while the second stage is lossless. Hence, for high-quality audio compression, the error induced by quantization, called *quantization noise*, must be kept below a certain tolerance level as determined by human perception. In this respect, recall from section 2.5 of Chapter 2 that the auditory system of the human ear can be modeled as the wavelet transform (with certain analyzing wavelet), at least as a first order of approximation. Hence, a suitable choice of the analyzing wavelet that matches the human audio perception certainly helps in the design of the quantizer.

In practice, the control of quantization noise below a tolerance (or noticeable) level is achieved through spectral analysis. The key feature in hearing from the spectral point of view is the masking property of certain dominant sounds over weaker ones in a neighborhood of a critical band as governed by some spreading function. For instance, a sinusoidal wave can be used as a mask. Hence, in the design of a wavelet audio compressor, a spectral analysis component is needed to help calculate the masking threshold. If a uniform quantizer is used, this thresholding parameter h is also used as the quantization step size as given by (7.2.2). Therefore, in view of (7.2.1) and Figure 7.8, the lossy component of the wavelet audio compressor can be summarized as in Figure 7.13. We remark that for stereo sound, a stereo control is used in addition to balancing the masking threshold calculation.

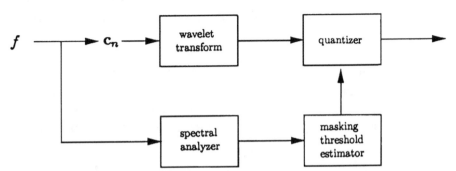

FIG. 7.13. *Lossy component of wavelet audio compressor.*

The most well known audio coding algorithm is MPEG audio, called masking-pattern universal subband integrated coding and multiplexing (MUSICAM). It is used in the moving pictures expert group MPEG-I standard.

The transform is a 32-band uniform filter bank (but not a wavelet transform in the strict sense), with an FFT as a spectral analyzer that facilitates the calculation of the masking curve. However, MUSICAM does not include an entropy encoder. It compresses audio signals of 700 Kbits/sec. (CD quality, i.e., 44.1 kHz sampling and 16 bits/sample) to about 128 Kbits/sec. for mono and 256 Kbits/sec. for stereo. Since the DWT decomposition algorithm can be realized as a two-band filter bank, multilevel decomposition or wavelet-packet decomposition leads to subband decomposition with "any" desirable number of bands and can be used in place of the filter bank.

7.2.3. Still image compression.

The most popular application of the DWT is in the area of image compression. Since we will be concerned only with the DWT or DWPT (discrete wavelet packet transform), the images to be compressed are assumed to be in digital format. State-of-the-art cameras are designed to produce digital images already. Otherwise, different kinds of digitizers are available. These include laser and drum scanners as well as solid-state cameras.

This subsection will focus on compression of still digital images. They are either in monochrome (sometimes called black and white) or in color. The resolution (or quality) of an image is measured by the number of pixels per square inch, and the image itself is represented with different numbers of bits per pixel. Hence, an 8-bit monochrome image means that the gray level for each pixel ranges from 0 to 255 (since $2^8 = 256$). On the other hand, to represent a color image with the same gray level, we need 24 bits, with 8 bits for red (R), 8 bits for green (G), and 8 bits for blue (B).

Let us first discuss wavelet compression of monochrome images. The domain of definition of such an image with 1024×512 resolution on a rectangular region $[a, b] \times [c, d]$, say, is given by the set of order pairs

$$\left(a + \frac{j}{2^{10}}(b - a), c + \frac{k}{2^9}(d - c) \right), \tag{7.2.13}$$

where $j = 0, \ldots, 2^{10} - 1$ and $k = 0, \ldots, 2^9 - 1$. Observe that $2^{10} = 1024$ and $2^9 = 512$. If this is an 8-bit monochrome image, then the range (or gray level) is given by

$$\{0, 1, \ldots, 255\}.$$

Of course a 1-bit image has range $\{0, 1\}$, a 4-bit image has range $\{0, \ldots, 15\}$, etc. Most black and white pictures are 8-bit images, but medical images are usually 12-bit. For convenience, we consider a change of variables in both the x- and y-directions, so that the domain described by (7.2.13) becomes

$$\left\{ \left(\frac{j}{2^m}, \frac{k}{2^n} \right) : 0 \le j \le 2^m - 1, \ 0 \le k \le 2^n - 1 \right\}, \tag{7.2.14}$$

and the image domain is the unit square $[0, 1]^2$. Hence, the input image data set for compression is given by

$$f\left(\frac{j}{2^m}, \frac{k}{2^n} \right) : 0 \le j \le 2^m - 1, \ 0 \le k \le 2^n - 1. \tag{7.2.15}$$

This puts us right back to the usual framework (see, for example, sections 6.1 and 6.4 in Chapter 6), except that the range of j and k in (7.2.15) are restricted. There are two ways to get around this restriction. One way is to use boundary scaling functions and wavelets as discussed in sections 6.6–6.7 in Chapter 6, but an easier and more efficient method is to create artificial data outside the unit square $[0, 1]^2$, simply by a reflection of the image across each of the four boundaries. We will take the second route. Since the scaling functions and wavelets for compression have compact support (or finite time duration), we will assume, without getting into much trouble, that the data set (7.2.15) is valid for all integers j and k. Consequently, the beginning of the discussion on two-dimensional feature extraction in subsection 7.1.5 and the entire subsection 7.2.1 on data compression schemes apply to our setting.

To compress a red, green, and blue (RGB) color image, one could compress each of the R, G, and B color components separately and put the compressed components back together in decompression. But to gain a much higher compression ratio, it is best to change RGB to luminancy and chrominancy. The luminance component (usually denoted by Y) is compressed exactly the same way as a monochrome image. On the other hand, the chrominance component can be compressed as much as 15 to 20 times the compression depth for the luminance component. Hence, by going from RGB to luminancy/chrominancy, one can increase the compression ratio by a factor of $2\frac{1}{2}$ or even higher.

For TV broadcast, the North American (i.e., NTSC) standard color space is YIQ, and the European one is YUV. Here, I, Q and U, V are linked to the chrominance component. The matrix to change from (U,V) to (I,Q) is given by

$$\begin{bmatrix} -0.2676 & 0.7361 \\ 0.3869 & 0.4596 \end{bmatrix}. \tag{7.2.16}$$

Finally, to change RGB to YUV, we have the linear transformation

$$\begin{bmatrix} 0.299 & 0.587 & 0.114 \\ -0.299 & -0.587 & 0.886 \\ 0.701 & -0.5887 & -.0114 \end{bmatrix} \tag{7.2.17}$$

if we follow the CCIR 601-1 recommendation, and the transformation

$$\begin{bmatrix} 0.2125 & 0.7154 & 0.0721 \\ -0.2125 & -0.7154 & 0.9279 \\ 0.7875 & -0.7154 & -0.0721 \end{bmatrix} \tag{7.2.18}$$

when the CCIR 709 recommendation is followed. To return to RGB from YUV, we simply multiply by the inverse of either the matrix (7.2.17) or the matrix (7.2.18). To go from RGB to YIQ, we first go to YUV and then change UV to IQ by multiplying by (7.2.16). To return from YIQ to RGB, we just take the inverse transforms. Of course, RGB and YIQ can be interchanged by a single matrix multiplication if (7.2.16) and (7.2.17) or (7.2.18) are combined in advance.

The DWT is a reversible operation, but quantization is a lossy compression scheme. Hence, to evaluate the performance of the lossy wavelet image compression algorithm, an objective measurement must be used. Currently, the most popular metric is the *peak signal-to-noise ratio* (PSNR), measured in db. It is defined by

$$\text{PSNR} = 20 \log_{10} \frac{2^b - 1}{\text{RMSE}}, \tag{7.2.19}$$

where b is the number of bits/pixel of the image (e.g., for an 8-bit image, $2^b - 1 = 2^8 - 1 = 255$). Also, RMSE is the root mean-square error of the compression defined by

$$\text{RMSE} = \left\{ \frac{1}{mn} \sum_{j=0}^{m-1} \sum_{k=0}^{n-1} \left[f\left(\frac{j}{2^m}, \frac{k}{2^n}\right) - g\left(\frac{j}{2^m}, \frac{k}{2^n}\right) \right]^2 \right\}^{1/2}, \tag{7.2.20}$$

where $f(x, y)$ is the original image with data set given by (7.2.15) and $g(x, y)$ is the reconstructed (compressed) image.

7.2.4. Video compression.

The reason that the DWT can be applied to achieve a very high degree of compression in lossy compression of still images is that it facilitates the effectiveness of quantization by exploiting the visually redundant information of a digital image. More precisely, both spatial and spectral redundant information can be exploited by the DWT. *Spatial redundancy* is a result of correlation of neighboring pixel values, and *spectral redundancy* is a consequence of correlation of spectral values (from more than one octave band) for the same pixel locations. In compression of images in a video sequence, there is more redundant information that can be exploited. With the exception of certain occasional scene changes, adjacent frames of a video sequence do not change by too much. This is called *temporal redundancy*. For instance, in a video sequence showing a lecture presented by an engineer, there is very little change between two consecutive frames. The only change is caused by the movement of the speaker (such as hand movement or motion of the lips). So, if $f_i(x, y)$ denotes the image of the ith frame, then there is very little content in the *difference image*

$$\delta_i(x, y) = f_{i+1}(x, y) - f_i(x, y), \tag{7.2.21}$$

as compared with that of the actual images $f_i(x, y)$ or $f_{i+1}(x, y)$. Consequently, the two images $f_i(x, y)$ and $\delta_i(x, y)$ together already provide a compression of the totality of the two image frames $f_i(x, y)$ and $f_{i+1}(x, y)$.

The content of the difference image $\delta_i(x, y)$ in (7.2.21) may be measured by its energy (i.e., the square of its L^2- or ℓ^2-norm). If there is absolutely no movement of the background scene and of the camera that captures the two image frames $f_i(x, y)$ and $f_{i+1}(x, y)$, then the energy of $\delta_i(, y)$ should be very small. Otherwise, the energy is more substantial. In the latter situation, an adjustment of the frame $f_{i+1}(x, y)$ is necessary. Usually, this adjustment only involves a simple shift, but a little change in size (magnification or contraction) may be necessary if the camera has moved backward or forward. If this change is substantial, it is more economical to consider it as a scene change.

When there is no scene change between the ith frame and the kth frame, where $k > i$, then the frames

$$f_i(x, y), \delta_i(x, y), \ldots, \delta_{k-1}(x, y) \tag{7.2.22}$$

can be used in place of the original frames

$$f_i(x, y), \ldots, f_k(x, y). \tag{7.2.23}$$

Observe that the frames in (7.2.23) can be perfectly reconstructed from those in (7.2.22) by consecutive addition; namely,

$$\begin{aligned} f_{i+1}(x, y) &= f_i(x, y) + \delta_i(x, y), \\ f_{i+2}(x, y) &= f_{i+1}(x, y) + \delta_{i+1}(x, y), \end{aligned} \tag{7.2.24}$$

$$\cdots \cdots \cdots \cdots \cdots$$

In addition, each of the frames in (7.2.22) can be considered as a still image itself, so that the wavelet compression scheme described in section 7.2.3 can be applied to compress these images. For example, to compress 3 to 4 seconds of a 24-bit RGB color video sequence (approximately 100 frames) without "dramatic" scene changes, if we assume a 70% saving in compressing the difference images in (7.2.22) over the original image sequence in (7.2.23), then it is possible to achieve a 1000 to 1 compression ratio or higher.

However, several technical difficulties must be overcome, among which is memory requirement. The major problem to solve is perhaps compensation for camera or scene movement in going from the image sequence (7.2.23) to the difference image sequences (7.2.22), as we pointed out earlier. To capture some video sequences, the camera is supposed to move. For instance, if the camera follows a moving vehicle, the background scene changes as well, though at a relatively slower speed, as compared with the video sequence rate of 30 frames/sec.

If an image sequence block (7.2.23) is fairly short, such as 12 to 15 frames, then the difficulties mentioned above are not serious. For instance, to solve the motion compensation problem, even the MPEG procedure could be followed. More precisely, the initial frame $f_i(x, y)$ that has been compressed can be treated as an I-frame (where "I" stands for intraframe compression). The wavelet compressed difference frames can be used, along with MPEG motion estimation and compensation, to give the predicted frames, called P-frames, by using (7.2.24). The B-frames can be produced by bidirectional interpolation between the I-frame and the first P-frame and between every two P-frames.

Another approach is to treat the video image sequence as a three-dimensional image and to apply three-dimensional wavelet data compression. If this approach is followed, the analyzing wavelet in the time direction should have very small support, since when long decomposition sequences are used, even moderate quantization tends to introduce high-frequency artifacts to the AC bands and motion deviations to the DC bands. A special feature of three-dimensional image compression is that if the time direction is considered as the z-axis, then the compressed three-dimensional image can be viewed not only along the z-direction but also along the x- and y-directions as well. Applications of this feature include viewing three-dimensional medical images and seismic blocks from all three directions.

7.3. Numerical solutions of integral equations.

Although the central theme of this book is "signal analysis," wavelets also have a great impact on other areas such as the numerical solution of differential and integral equations. In fact, the notion of multiresolution analysis (MRA) is similar to the multilevel approach in the solution of partial differential equations (PDEs), and the concept of wavelet bases is analogous to that of hierarchical bases for the multigrid method. The main difference is that since almost all differential and integral equations of interest are formulated on domains with a nontrivial boundary, the operations of translation and dilation of certain basic functions (such as wavelets) alone cannot take care of the required boundary conditions. For this type of application, we could use boundary scaling functions and boundary wavelets, such as those introduced in sections 6.6–6.7 in Chapter 6. It is important to point out that the multilevel wavelet basis functions generated by the boundary (and interior) wavelets retain the zoom-in and zoom-out capability as well as the property of vanishing moments relative to the (bounded or one-sided bounded) intervals. For the multivariate setting, we can simply use the tensor products (see (6.4.6), (6.4.12)–(6.4.14), and (6.4.16)) of both interior and boundary scaling functions and wavelets (see sections 6.6–6.7).

Since much more has been written on the wavelet solution of PDEs than that of integral equations, we choose to discuss the second topic. In the following, we will only consider the wavelet solution of integral equations of the first kind. It should be emphasized that the same approach also applies to other types of differential or integral equations.

The advantages of using wavelet basis functions for the solution of integral equations include (i) flexibility in choosing the appropriate scaling functions and corresponding wavelets for better waveform matching, (ii) solution segmentation for localized analysis, (iii) vanishing moments resulting in very sparse coefficient matrices after thresholding, and (iv) computational efficiency.

7.3.1. Statement of problem.

The objective of this section is to describe an efficient solution of the class of integral equations

$$\int_a^b f(y)K(x,y)dy = g(x), \tag{7.3.1}$$

where $f(y)$ is an unknown function, but the kernel $K(x,y)$ and the function $g(x)$ are given. This equation, depending on the kernel and the limits of integration, is referred to by different names such as Fredholm, Volterra, or Weiner–Hopf integral equations. This type of integral equations appear frequently in practice such as in inverse problems where the goal is to reconstruct the function $f(x)$ from a set of known data represented in the functional form of $g(x)$. For instance, in some electromagnetic scattering problems to be discussed later, the current distribution on the metallic surface is related to the incident field in the form of an integral equation of type (7.3.1) with Green's function as the integral kernel. Observe that solving for $f(x)$ is equivalent to finding the inverse transform of $g(x)$ with respect to the kernel $K(x, y)$. In particular, if $K(x, y) = e^{jxy}$, then $f(x)$ is nothing but the inverse Fourier transform of $g(x)$. We will assume that (7.3.1) has a unique solution. Although we consider only solutions of integral equations of the first kind, we remark that the method can be extended to second-kind and higher-dimension integral equations with little additional work.

In the conventional method of moments (MoMs), the boundary of the domain of integration is approximated by discretization into several segments. The unknown function is then expanded in terms of certain known basis functions with unknown coefficients, which may occupy more than one segment. Finally, the resultant equation is tested with the same or different functions resulting in a set of linear equations whose solution gives the unknown coefficients. Because of the nature of the integral operator, the matrix is usually dense, and the inversion and final solution of such a system of linear equations is very time consuming, particularly for solving scattering problems that involve electrically large objects. Consequently, the use of conventional MoMs in solving integral equations is generally restricted to low-frequency cases.

In order to overcome some of the difficulties such as large memory requirement and high computation time for inverting the dense matrix associated with the conventional methods for solving integral equations, it is natural to consider using certain wavelet basis functions. It should be pointed out, however, that even with a wavelet basis, the resultant matrix is still dense, as it should be because of the very nature of the integral operator; however, primarily because of the small local supports and vanishing moment properties, many of the matrix elements are very small as compared with the largest element and, hence, can be dropped without significantly affecting the solution. In addition, unlike the conventional MoM, the wavelet MoM has a multilevel nature in the sense that the domain of integration is essentially discretized with different discretization steps corresponding to different levels or scales.

However, the difficulty with using wavelets is that the boundary conditions need to be enforced explicitly. In the following, we will apply compactly supported semiorthogonal spline wavelets specially constructed for the bounded interval introduced in sections 6.6–6.7 for solving boundary integral equations. We have chosen the semiorthogonal wavelets primarily because of the following reasons.

(i) Unlike most of the orthonormal wavelets, compactly supported semiorthogonal spline wavelets have closed-form expressions.

(ii) Unlike the nonorthogonal (or biorthogonal) wavelets, semiorthogonal wavelets allow the change of bases (see (7.1.8)–(7.1.11) for the interior ones).

(iii) These wavelets are symmetric and hence have a generalized linear phase, an important factor for reconstructing the function.

(iv) Because of the "total positivity" properties of splines, they have certain very desirable properties for waveform matching (see section 5.4 in Chapter 5).

7.3.2. Boundary B-splines and B-wavelets.

We will consider knot sequences

$$\mathbf{t}^s = \{t_k^s\}_{k=-m+1}^{2^s+m-1}$$

with

$$\begin{cases} t_{-m+1}^s = t_{-m+2}^s = \cdots = t_0^s = 0, \\ t_k^s = \dfrac{k}{2^s}, \quad k = 1, \ldots, 2^s - 1, \\ t_{2^s}^s = t_{2^s+1}^s = \cdots = t_{2^s+m-1}^s = 1, \end{cases}$$

as in Example 6.14 (after an appropriate change of scale from $[0, n]$ to $[0, 1]$).

To ensure that there are interior B-splines and B-wavelets, we always consider $2^s \geq 2m - 1$ and let s_0 be such an s. Then for each $s \geq s_0$, we have the scaling functions

$$\phi_{m,s,k}(x) := \begin{cases} N_{\mathbf{t}^s,m;k}(2^{s-s_0}x), & k = -m+1, \ldots, -1, \\ N_m(2^s x - k), & k = 0, \ldots, 2^s - m, \\ N_{\mathbf{t}^s,m;k}(1 - 2^{s-s_0}x), & k = 2^s - m + 1, \ldots, 2^s - 1 \end{cases}$$

(where $N_{\mathbf{t}^s,m;k}(x)$ are the boundary B-splines in (6.6.11) and Examples 6.11–6.13, and $N_m(x)$ is the mth-order cardinal B-spline with integer knots) and the B-wavelets

$$\psi_{m,s,k}(x) := \begin{cases} \psi_{\mathbf{t}^{s-1},\mathbf{t}^s,m;k}(2^{s-s_0}x), & k = -m+1, \ldots, -1, \\ \psi_m(2^s x - k), & k = 0, \ldots, 2^s - m + 1, \\ \psi_{\mathbf{t}^{s-1},\mathbf{t}^s,m;k}(1 - 2^{s-s_0}x), & k = 2^s - 2m + 2, \ldots, 2^s - m \end{cases}$$

(where $\psi_{\mathbf{t}^{s-1},\mathbf{t}^s,m;k}(x)$ are the boundary wavelets introduced in section 6.7 and $\psi_m(t)$ is the mth-order (cardinal) B-wavelet introduced in (5.2.24)–(5.2.25) of Chapter 5).

In what follows, we will omit the subscript m in $\phi_{m,s,k}(x)$ and $\psi_{m,s,k}(x)$ if the order m is fixed and is understood.

7.3.3. Solution of integral equations using boundary spline wavelets.

In this subsection, we will show the advantage of the scaling functions $\phi_{s,k}(x) = \phi_{m,s,k}(x)$ and wavelets $\psi_{s,k}(x) = \psi_{m,s,k}(x)$ introduced above for solving integral equations of the form given by (7.3.1).

With the domain of integration being $[0, 1]$, let us expand the unknown function $f(y)$ in (7.3.1) in terms of the scaling functions and wavelets on the bounded interval as

$$f(y) = \sum_{k=-m+1}^{2^{s_u+1}-1} c_{s_u,k}\phi_{s_u,k}(y) \tag{7.3.2}$$

$$= \sum_{s=s_0}^{s_u} \sum_{k=-m+1}^{2^s-m} d_{s,k}\psi_{s,k}(y) + \sum_{k=-m+1}^{2^{s_0}-1} c_{s_0,k}\phi_{s_0,k}(y), \tag{7.3.3}$$

where the last equality is the consequence of the MRA properties of the scaling function, by virtue of which we can write

$$V_{s+1} = W_s \oplus V_s. \tag{7.3.4}$$

A telescopic expansion of (7.3.4) starting with $s = s_u$ and ending with $s = s_0$ yields (7.3.2). It should be pointed out here that the wavelets $\{\psi_{s,k}(x)\}$ by themselves form a complete set; therefore, the unknown function could be expanded entirely in terms of the wavelets. However, to retain only a finite number of terms in the expansion, the scaling function part in (7.3.3) must be included. In other words, because of their bandpass filter characteristics, the wavelets $\psi_{s,k}(x)$ extract successively lower and lower frequency components of the unknown function with decreasing values of the scale parameter s, while $\phi_{s_0,k}(x)$, because of its lowpass filter characteristics, retains the lowest frequency components or the coarsest approximation of the original function.

The second expansion of $f(y)$ given by (7.3.2) is substituted in (7.3.1), and the resultant equation is tested with the same set of expansion functions. This results in the so-called Petrov–Galerkin method which gives a set of linear equations

$$\begin{bmatrix} [A_{\phi,\phi}] & [A_{\phi,\psi}] \\ [A_{\psi,\phi}] & [A_{\psi,\psi}] \end{bmatrix} \begin{bmatrix} [c_{s_0,k}]_k \\ [d_{s,n}]_{s,n} \end{bmatrix} = \begin{bmatrix} \langle E_z^i, \phi_{s_0,k'}\rangle_{k'} \\ \langle E_z^i, \psi_{s',n'}\rangle_{s',n'} \end{bmatrix}, \tag{7.3.5}$$

where

$$[A_{\phi,\phi}] := \langle \phi_{s_0,k'}, (L_K\phi_{s_0,k})\rangle_{k,k'}, \tag{7.3.6}$$

$$[A_{\phi,\psi}] := \langle \phi_{s_0,k'}, (L_K\psi_{s,n})\rangle_{k',s,n}, \tag{7.3.7}$$

$$[A_{\psi,\phi}] := \langle \psi_{s',n'}, (L_K\phi_{s_0,k})\rangle_{k,s',n'}, \tag{7.3.8}$$

$$[A_{\psi,\psi}] := \langle \psi_{s',n'}, (L_K\psi_{s,n})\rangle_{s,n,s',n'}, \tag{7.3.9}$$

$$\langle f, g \rangle := \int_0^1 f(x)g(x)dx, \tag{7.3.10}$$

$$(L_K f)(x) := \int_0^1 f(y)K(x,y)dy, \tag{7.3.11}$$

and the subscripts k, k', s, n, s', n' satisfy

$$k, k' = -m + 1, \ldots, 2^{s_0} - 1,$$

$$s, s' = s_0, \ldots, s_u,$$

$$n = -m + 1, \ldots, 2^s - m,$$

$$n' = -m + 1, \ldots, 2^{s'} - m.$$

In (7.3.5), $[d_{s,n}]_{s,n}$ is a vector and should not be confused with a matrix. Here, the index n is varied first for a fixed value of s. It can be shown that the total number N of unknowns in (7.3.5), interestingly, does not depend on s_0 as given by

$$N = 2^{s_u+1} + m - 1.$$

A few explanations of (7.3.2)–(7.3.3) and (7.3.5) are in order. Observe that in conventional MoMs, we use the representation of f similar to (7.3.2) in terms of $\{\phi_{s,k}(x)\}$ only. With this expansion the total number of unknowns will be 2^{s_u+1}; therefore, in wavelet method, there is an increase of $m-1$ unknowns (1 for the linear case and 3 for the cubic case), which is usually insignificant when large problems are to be solved. Even though the limits of integrations in (7.3.9) and (7.3.10) range from 0 to 1, the actual integration limits are much smaller because of the finite supports of the semiorthogonal scaling functions and wavelets.

We can explain the "denseness" of the conventional MoM and the "sparseness" of the wavelet MoM by recalling the fact that unlike wavelets, the scaling functions do not have the vanishing moments properties. Consequently, for two pulse or triangular functions $\phi_1(x)$ and $\phi_2(x)$ (usual bases for the conventional MoM and suitable candidates for the scaling functions), even though $\langle \phi_1, \phi_2 \rangle = 0$ for nonoverlapping supports, the quantity $\langle \phi_1, L_K \phi_2 \rangle$ is not small, since $(L_k \phi_2)(x)$ is not small. On the other hand, because of the compact support of the wavelets, the integral vanishes if the function against which the wavelet is being integrated behaves as a polynomial of a certain order locally. Consequently, the integrals such as $(L_K \psi_{s,n})$ and the inner products involving the wavelets are very small for nonoverlapping supports. (See section 6.5 for a more in-depth explanation.)

From the definition of B-splines, the scaling function has the "smoothing" effect on a function against which it is integrated. The smoothing effect can be understood as follows. If we convolve two pulse functions, both of which are discontinuous but totally positive, the resultant function is a linear spline that is continuous. Likewise, if we convolve two linear B-splines, we get a cubic B-spline that is twice continuously differentiable. Analogous to these, the function $L_K \phi_{s_0,k}$ is smoother than the kernel K itself. Furthermore, because of the semiorthogonality of the wavelets, we see that

$$\langle \phi_{s,k}, \psi_{s',\ell} \rangle = 0, \ s \leq s',$$

and that the integrals $\langle \phi_{s_0,k'}, (L_K \psi_{s,n}) \rangle$ and $\langle \psi_{s',n'}, (L_K \phi_{s_0,k}) \rangle$ are quite small.

The $[A_{\phi,\phi}]$ portion of the matrix, although diagonally dominant, does not usually have very small entries when compared with the diagonal entries. As mentioned before, in the case of conventional MoMs, all the elements of the matrix are of the form $\langle \phi_{s,k'}, (L_K \phi_{s,k}) \rangle$. Consequently, we cannot sparsify the matrix by thresholding. In the case of wavelet MoMs, the entries of $[A_{\phi,\phi}]$ occupy a very small portion (5×5 for the linear setting and 11×11 for the cubic spline setting) of the matrix, while the rest consists of entries whose magnitudes are very small as compared with the largest entry; hence, a significant number of entries can be set to zero without affecting the solution appreciably.

7.3.4. Sparsity and error considerations.

The study of the effects from thresholding the matrix elements that results in sparsity and error in the solution is the objective of this subsection. By thresholding, we mean, as in the feature extraction procedure, setting to zero those elements of the matrix that are smaller (in magnitude) than some positive number δ, called the thresholding parameter, times the largest element of the matrix.

Let $A_{\max}$ and $A_{\min}$ be the largest and smallest elements of the matrix in (7.3.5). For a fixed value of the threshold parameter δ, we define % relative error (ϵ_δ) as

$$\epsilon_\delta := \frac{\|f_0 - f_\delta\|_2}{\|f_0\|_2} \times 100 \qquad (7.3.12)$$

and % sparsity (S_δ) as

$$S_\delta := \frac{N_0 - N_\delta}{N_0} \times 100.$$

In the above, f_δ represents the solution obtained from (7.3.5), when the elements whose magnitudes are smaller than $\delta A_{\max}$ have been set to zero. Similarly, N_δ is the total number of elements left after thresholding. Clearly, $f_0(x) = f(x)$ and $N_0 = N^2$, where N is the number of unknowns.

Table 7.1 gives an idea of the relative magnitudes of the largest and the smallest elements in the matrix for conventional and wavelet MoMs. As is expected, because of their higher vanishing moment property, cubic spline wavelets give the higher ratio, $A_{\max}/A_{\min}$.

TABLE 7.1. *Relative magnitudes of the largest and the smallest elements of the matrix for conventional and wavelet MoMs.* $a = 0.12\delta_0$.

	Conventional MoM	Wavelet MoM ($m = 2$)	Wavelet MoM ($m = 4$)
$A_{\max}$	5.377	0.750	0.216
$A_{\min}$	1.682	7.684×10^{-8}	8.585×10^{-13}
Ratio	3.400	9.761×10^6	2.516×10^{11}

With the assumption that the $[A_{\phi,\phi}]$ part of the matrix is unaffected by the thresholding operation, which is a fairly reasonable assumption, it can be shown that

$$S_\delta \leq \left[1 - \frac{1}{N} - \frac{(2^{s_0} + m - 1)(2^{s_0} + m - 2)}{N^2}\right] \times 100 . \qquad (7.3.13)$$

As mentioned before, the total number of unknowns is independent of s_0, the lowest level of discretization. However, it is clear from (7.3.12) that the upper limit of S_δ increases with the decreasing value of s_0. Therefore, it is better to choose $s_0 = \lceil \log_2(2m - 1) \rceil$, where $\lceil x \rceil$ represents the smallest integer that is greater than or equal to x.

7.3.5. Numerical examples.

In this section we present some numerical examples for the problem that involves an infinitely long metallic cylinder illuminated by a TM plane wave. The z-directed surface current (J_{sz}) distribution on the cylinder is related to the incident field and Green's function by an integral equation

$$j\omega\mu_0 \int_C J_{sz}(\ell')G(\ell, \ell')d\ell' = E_z^i(\ell), \tag{7.3.14}$$

where

$$G(\ell, \ell') = \frac{1}{4j}H_0^{(2)}(k_0|\rho(\ell) - \rho(\ell')|),$$

and E_z^i is the z-component of the incident electric field. Here, the contour of integration has been parameterized with respect to the chord length. It is clear that (7.3.14) is of the form of (7.3.1). Our objective is to solve (7.3.14) for the unknown current distribution J_{sz}.

The matrix equation (7.3.5) is solved for a circular cylindrical surface. Figures 7.14–7.15 show the surface current distribution for the linear spline case with different sizes of the cylinder. Results for the cubic spline case are found to be in very good agreement. Wavelet MoM results are compared with the conventional MoM results. To obtain the conventional MoM results, we may use triangular functions for expanding both the unknown current distribution and for testing the resultant equation. The conventional MoM results have been verified with the series solution.

Again, the results of the conventional MoM and the wavelet MoM agree very well. The matrix sizes for the conventional, linear wavelet, and cubic wavelet MoMs are 32×32, 33×33, and 35×35, respectively.

Effects of δ on the error in the solution and the sparsity of the matrix are shown in Figure 7.16. The magnitude of the error increases rapidly for the linear spline case. The matrix elements with $\delta = 0.0002$ for the linear spline case are shown in Figure 7.17. A darker color on an element indicates a larger magnitude. Figure 7.18 gives an idea of the pointwise error in the solution for the linear case. In Figure 7.19, we present the thresholded matrix $(\delta = 0.0025)$ for the cubic spline case. The $[A_{\psi,\psi}]$ part of the matrix is almost diagonalized. Finally, in Figure 7.20 we can see that even with the matrix as shown in Figure 7.19, we get a very good solution for the current distribution.

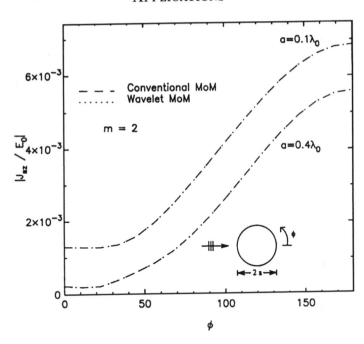

FIG. 7.14.   *Magnitude of the surface current distribution on a metallic circular cylinder computed using linear spline wavelet MoMs and conventional MoMs.*

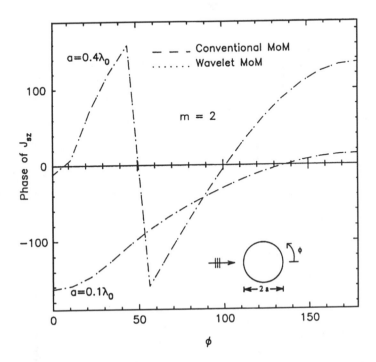

FIG. 7.15.   *Phase of the surface current distribution on a metallic circular cylinder computed using linear spline wavelet MoMs and conventional MoMs.*

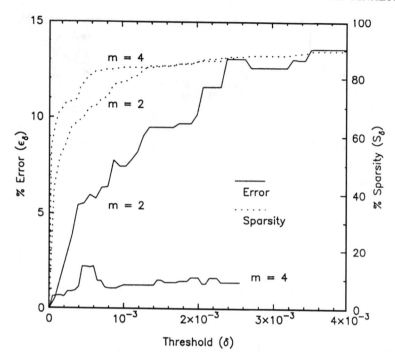

FIG. 7.16.　*Sparsity of the matrix and the error in the solution of the surface current distribution as a function of the threshold parameter using wavelet MoMs.*

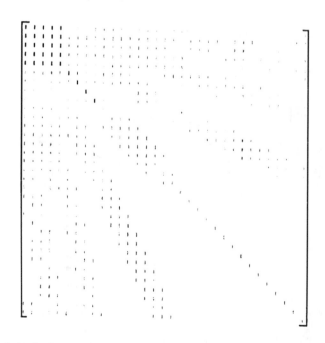

FIG. 7.17.　*A typical gray-scale plot of the matrix elements obtained using linear wavelet MoMs with* $\delta = 0.0002$.

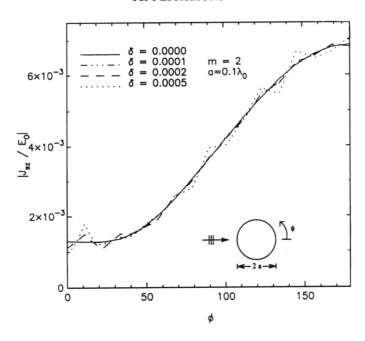

FIG. 7.18.   *Magnitude of the surface current distribution computed using linear wavelet MoMs for different values of the thresholding parameter δ.*

FIG. 7.19.   *A typical gray-scale plot of the matrix elements obtained using cubic wavelet MoMs with δ = 0.0025.*

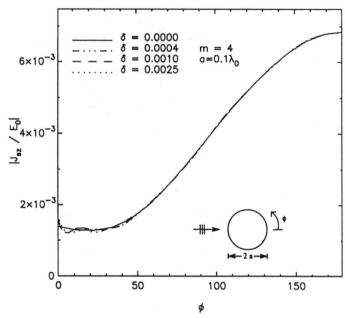

FIG. 7.20.   *Magnitude of the surface current distribution computed using cubic wavelet MoMs for different values of the thresholding parameter δ.*

It is worth pointing out here that no matter what the size of the matrix is, only a 5×5 block in the case of linear splines and an 11×11 block in the case of cubic splines (see the top-left corners of Figures 7.17 and 7.19) will remain unaffected by thresholding; a significant number of the remaining elements can be set to zero without causing much error in the solution.

Summary and Notes

Chapter 1.

The objective of this chapter is to give an overview of wavelets and to clarify the concept of the wavelet transform as a mathematical tool for signal analysis. Four well-known classes of basic functions (trigonometric functions, Walsh functions, splines, and the Haar function) are used to bring out the four major ingredients of the wavelet transform as a mathematical tool for signal analysis, namely, waveform matching, signal segmentation, time-frequency localization with the zoom-in and zoom-out capability, as well as computational and implementational efficiency. Although it is clear that the Walsh series is favored over the trigonometric (i.e., sine and/or cosine) series for matching a "square wave," the importance of the Walsh basis functions introduced by Walsh in [VI.42] is that their segmentations by means of the first-order cardinal B-spline $\phi(t)$ (see Schoenberg [III.5]) and the Haar function $\psi_H(t)$ (introduced by Haar in [VI.23]) give rise to the notions of multiresolution analysis (MRA), introduced by Meyer [VI.30] and Mallat [VI.29], and the discrete wavelet transform (DWT) as coefficients of the (orthonormal) Haar series expansion. Furthermore, computation of the DWT values is accomplished simply by the so-called wavelet decomposition algorithm, usually called the Mallat algorithm (see [VI.29]), although it was already used for image coding by Burt and Adelson in [VI.5]. The DWT is a doubly indexed sequence of discrete samples of the integral wavelet transform (IWT), also known as the continuous wavelet transform (CWT), introduced by Grossmann and Morlet in [VI.22]. Hence, these DWT values give localized time-frequency information with the zoom-in and zoom-out capability. When this concept is discussed in section 1.2, the notion of dual wavelets is introduced to allow more flexibility for wavelet filter design. Here, "duality" is equivalent to the concept of biorthogonal wavelet bases considered by Cohen, Daubechies, and Feauveau [VI.14] for constructing compactly supported and symmetric or antisymmetric wavelets as well as by the author and Wang [VI.11] for constructing semiorthogonal compactly supported spline wavelets. It is also important to point out that the Mallat algorithm is also used in subband coding in digital signal processing using filter banks (see, for example,

[VI.40], [VI.41], and the recent monograph [I.6, I.7, I.8, I.10]). The important difference is that instead of processing a digital signal (i.e., digital samples of an analog signal), the wavelet decomposition/reconstruction algorithm is used to process the coefficient sequence of the series expansion of the analog signal representation in terms of certain translates of a given scaling function with a given scale by some integer power of 2.

Chapter 2.

This chapter is devoted to a study of time-frequency localization. The standard approach in ideal lowpass and bandpass filtering for separating an analog signal into different frequency bands is used to motivate the need of time localization. Hence, the notion of root mean square (RMS) bandwidth is introduced (to replace the classical bandwidth) in terms of "centers" and "widths," which follow from the notion of standard deviation used for statistical measurement. A function with both finite RMS duration and finite RMS bandwidth is used as a window function to define the short-time Fourier transform (STFT). The first paper that introduced the STFT is Gabor [VI.20], where a Gaussian function was used as the window function. Hence, the STFT is also called the Gabor transform. The uncertainty principle (see, for example, [I.1]) says that the Gaussian functions are the only ones that achieve the uncertainty lower bound $\frac{1}{2}$. It is shown in section 2.3 that the window size of the STFT is rigid and, hence, not quite suitable both for detecting sharp changes and for studying low-frequency environments. The width of the window in the integral wavelet transform (IWT), on the other hand, adjusts itself automatically according to the change in frequency. This is done by picking the scale value to align with some known reference frequency value and letting the scale parameter vary in inverse proportion to the frequency variable. The one-sided center and radius of the Fourier transform of this (bandpass) window function are introduced in order for the frequency window to slide along the (positive) frequency axis, and, hence, a new uncertainty principle (see [VI.13], for example) is needed. It turns out that it has the same uncertainty lower bound of $\frac{1}{2}$ that, however, cannot be attained. This chapter ends with a discussion of modeling the cochlea of the human ear. By using the assumption that the spiral geometry of the cochlea induces a logarithmic delay in frequency, it is shown that the reception at the sensory cell located at the position x is given by the IWT with scale e^{-x}, with the exception that the square root is not taken for the normalization in the definition of the IWT. The logarithmic delay assumption happens to agree with an empirical result of Zweig in [VI.43].

Chapter 3.

The notion of multiresolution analysis (MRA) of Mallat and Meyer provides a very powerful tool for the construction of wavelets and implementation of the wavelet decomposition/reconstruction algorithms. The sampling theorem is used to motivate the formulation of analog signal representations in terms of superpositions of certain uniform shifts of a single function called a scaling function. Stability of this signal representation is achieved by imposing the Riesz condition on the scaling function, which, in turn, is one of the main ingredients in the definition of an MRA. Another important condition of an MRA is the nested sequence of subspaces as a result of using scales by integer powers of 2. Two methods for mapping a signal to an MRA subspace discussed in section 3.1 are orthogonal projection and interpolation of discrete data.

When the Shannon sampling function is used as a scaling function, the digital samples (at uniform sample sites) of bandlimited signals with sufficiently small bandwidth agree with the coefficients obtained by orthogonal projection. (See, for example, [I.9] for further observations in this direction.) If the first-order B-spline is used as a scaling function, then the same digital samples also agree with the coefficients obtained by interpolation at the uniform sample sites. However, since the Shannon sampling function has infinite root mean square (RMS) duration, while the first-order B-spline has infinite RMS bandwidth, we need other scaling functions in order to generate the corresponding wavelets for better time-frequency localization. Higher-order cardinal B-splines are introduced for this purpose. It turns out that as the order tends to infinity, the uncertainty lower bound $\frac{1}{2}$ is achieved as the limiting value (see [VI.13] and Unser, Aldroubi, and Eden [VI.39], where it is proved that certain scales of the mth-order cardinal B-splines converge in L^2 to a Gaussian function as m tends to infinity). Furthermore, not only do the B-splines have the bell (or Gaussian) shape, the magnitudes of their Fourier transform are almost bell shaped and, consequently, have very small side lobes. Two measurements of the side-lobe to main-lobe ratios are introduced in section 3.4. Finally, a summary of the nice properties of the cardinal B-splines is included at the end of the chapter. More details can be found in [I.1] and [III.2–III.6].

Chapter 4.

Orthonormal wavelets are easier to appreciate, since an orthonormal wavelet series is nothing but a generalized Fourier series whose coefficients constitute the discrete wavelet transform (DWT) of the function represented by this series. However, since an orthonormal wavelet is self-dual, the same wavelet is used both as an analyzing wavelet (to define the integral wavelet transform (IWT)) and as a synthesizing wavelet (to match the waveform of the signal). In this chapter, three approaches are used to construct orthonormal wavelets. The first approach is to start with a known scaling function such as a cardinal B-spline or a fundamental interpolating spline. In the case of cardinal B-splines, an orthonormalization process is used to produce an orthonormal scaling function and, hence, its corresponding orthonormal wavelet by a suitable modification of the two-scale sequence. The orthonormalization process was introduced by Schweinler and Wigner [VI.34], and the resulting wavelets are the Battle–Lemarié wavelets, obtained independently by Battle [VI.3] and Lemarié [VI.27], using different methods. When the integer knots are used on the negative t-axis and half-integer knots are used for the positive t-axis to define the fundamental interpolating splines of order $2m$, then the mth derivatives evaluated at $2t - 1$ are constant multiples of the Strömberg wavelets [VI.38]. This was proved by Lai in [VI.25]. The second approach is to start with a desirable shape of the Fourier transform of an orthonormal scaling function. This is analogous to lowpass filter design in classical signal processing. For instance, the orthonormal Meyer scaling functions introduced by Meyer in [VI.31] are constructed by smoothing the jump discontinuities of the filter characteristic of the ideal lowpass filter with passband from $-\pi$ to π. The orthonormality property is guaranteed by the condition that the sum of the squares of the $2\pi k$ translates (for all integers k) of this magnitude spectrum is equal to 1. The construction also requires the 4π-periodic extension of this filter characteristic to be computable in order to give the two-scale symbol of the scaling function. The third approach in constructing orthonormal wavelets considered in Chapter 4 is to construct a two-scale symbol directly under the quadrature mirror filter (QMF) constraint as described by equation (4.49). This route was taken by Daubechies in [VI.17], where the binomial two-scale symbol of the mth-order cardinal B-spline is used as a factor of the two-scale symbol to ensure convergence of certain infinite products and to maintain the m vanishing moments. Phase retrieval of the other polynomial factor is achieved by choosing an appropriate half of the roots of the reciprocal polynomial with real coefficients, formulated by taking the product of a monomial (to change a Laurent polynomial to an algebraic one) and the square of the absolute value of this polynomial factor.

Chapter 5.

The only orthonormal wavelet that is symmetric or antisymmetric (for linear-phase or generalized linear-phase filtering) and has compact support (to give finite decomposition and reconstruction sequences) is the "Haar wavelet." This was proved by Daubechies [VI.17]. Observe that when an orthonormal wavelet is used as the analyzing wavelet, it is used as the synthesizing wavelet as well. The objective of this chapter is to introduce a dual pair of wavelets to share the workload. The formulation of duality in section 5.1 follows the framework of biorthogonal wavelets introduced by Cohen, Daubechies, and Feauveau [VI.14], where symmetry (or antisymmetry), without sacrificing finite filtering, is achieved by giving up orthogonality in decomposition. However, without this orthogonality requirement in decomposition, duality does not allow interchanging the dual pair when the wavelet decomposition/reconstruction scheme is only used for a finite number of levels. In other words, orthogonal decomposition requires the notion of semiorthogonal wavelets introduced by the author in [I.1]. Here, semiorthogonality means that orthogonality is required among different levels but is not required for translations. The first semiorthogonal wavelets were spline wavelets with minimum support (or B-wavelets) constructed by the author and Wang in [VI.11], independently (for even orders) by Aldroubi and Unser (see [IV.2, pp. 91–122]), and in Fourier transform formulation by Auscher [VI.2]. The duality principle that allows interchanging the dual pair was first introduced in [VI.11].

Another reason for giving up orthogonality in translation is that high-order orthonormal wavelets have poor time-frequency localization. In fact, the filter characteristics of the Daubechies wavelets tend to that of an ideal bandpass filter, and the area of the time-frequency window tends to infinity as the order increases. This result was established by the author and Wang in [VI.12]. The important concept that gives rise to asymptotic optimal time-frequency localization is total positivity. This observation and other related results along this line are due to the author and Wang in [VI.13]. It was also shown by Unser, Aldroubi, and Eden [VI.39] that certain scaling functions and their corresponding semiorthogonal wavelets that are results of m-fold convolutions also tend to the Gaussian and modulated Gaussian, respectively, as m tends to infinity. The notions of stoplets and cowlets were introduced in [VI.13]. The most comprehensive writing on total positivity is by Karlin [III.3].

Chapter 6.

Wavelets are attractive mathematical tools because of their efficient algorithms and computational schemes. This chapter is devoted to algorithm development and a study of boundary wavelets. Section 6.2 is concerned with analog signal representation in a multiresolution analysis (MRA) subspace. In particular, several interpolation schemes are discussed. Polynomial reproduction is emphasized to utilize the vanishing-moment property of the integral wavelet transform (IWT). When fast algorithms are needed, quasi interpolation is favored over discrete data interpolation. The quasi-interpolation scheme in terms of partial sums of a Neumann series expansion was introduced by the author and Diamond in [VI.7]. Only two terms are used in Example 6.5. When a better interpolation result is required, one can use more terms of the Neumann series. The general framework of orthogonal decompositions and reconstructions in section 6.2 was introduced by the author and De Villiers in [VI.6] for the purpose of constructing the decomposition matrices and dual wavelets for spline wavelets with arbitrary knots on a bounded interval. A general framework of multidimensional wavelets with matrix dilation was developed by the author and Li in [VI.8], but the emphasis in section 6.4 is on multifrequency wavelet transform. This notion was introduced by the author and Shi in [VI.10], where a general algorithm is given. Examples of various decompositions are shown in Figures 6.2–6.6, which include the wavelet packet decomposition of Coifman, Meyer, Quake, and Wickerhauser [VI.16]. The fast algorithms from Beylkin, Coifman, and Rokhlin [VI.4] can be used along this line. Orthogonal and biorthogonal boundary wavelets as well as their decomposition and reconstruction matrices are given by Andersson, Hall, Jawerth, and Peters [VI.1], Cohen, Daubechies, and Vial [VI.15], and others. Our emphasis, however, is on spline wavelets, since we allow arbitrary knots for analyzing discrete data on nonuniform samples. The first boundary spline-wavelet paper is [VI.9] by the author and Quak. The corresponding filter matrices were given by Quak and Weyrich [VI.32]. Only uniform knots were allowed in these earlier papers. On the other hand, although the nonuniform knot setting was already considered by Lyche and Mørken [VI.28], there were no results on wavelet decomposition, duals, etc. in the literature, until the paper by the author and De Villiers [VI.6] was completed. However, since the construction in [VI.6] is relatively more complicated, we use a more direct (but somewhat more computationally expensive) scheme in section 6.7 by following the first approach given in [VI.28]. The recursive algorithm for computing the P_m matrices was also developed in [VI.6]. Being a variation of the so-called Oslo algorithm, it is more efficient than a direct application of the de Boor and Fix quasi-interpolation formula (see [III.1, pp. 116–117]).

Chapter 7.

This final chapter highlights some of the most common applications of the wavelet transform. By taking full advantage of the properties of vanishing moments and small compact support (or finite time duration) of the analyzing wavelet, in section 7.1 we first apply the wavelet transform to the detection of isolated singularities for one-dimensional signals and feature extraction for higher-dimensional signals such as images. Here, the method to compensate for the translation variance of the discrete wavelet transform (DWT) is due to Shensa [VI.36]. Sections 7.1.4–7.1.5 that describe a spline-wavelet detection procedure and feature extraction also serve as preparation material for the presentation of wavelet data compression.

A fairly complete treatment of the wavelet data compression procedure is given in subsection 7.2.1. It includes a discussion of the procedures of quantization and of certain entropy coding schemes. The reader is referred to the standard texts [II.1–6] and research articles [VI.24, VI.37] for an in-depth treatment of these two topics. For image compression, the embedded zero-tree due to Shapiro [VI.35] and its improvement due to Said and Pearlman [VI.33] are particularly effective. For audio compression discussed in subsection 7.2.2, the reader is referred to the articles in [VI.18] for supplementary study. Wavelet compression of still images and video image sequences are studied in subsections 7.2.3 and 7.2.4, respectively. More details on the color transforms given in (7.2.16)–(7.2.18) can be found in [II.6], and other related but somewhat different wavelet-based video compression schemes can be found in [I.7] and [VI.19]. The video compression standards for video broadcasting are MPEG-I and MPEG-II. MPEG stands for the moving pictures experts group of the International Standardization Organization (ISO). MPEG-I is a medium quality compression standard and MPEG-II a higher-quality standard. The reader is referred to [II.5] and [VI.26] for more details.

Although the main theme of wavelet applications in this monograph is signal analysis, wavelets can also be used very effectively as basis functions in numerical solutions of differential and integral equations. Details of this approach can be found in the fundamental paper [VI.4] of Beylkin, Coifman, and Rokhlin. We choose integral equations as an example, since most writings in the literature are concerned with differential equations (see [V.1]). The presentation in section 7.3 is a summary of the joint paper of Goswami, Chan, and the author [VI.21].

References

I. Related Wavelet Books in English.

1. C. K. Chui, *An Introduction to Wavelets*, Academic Press, Boston, 1992.
2. I. Daubechies, *Ten Lectures on Wavelets*, CBMS Series 61, SIAM, Philadelphia, 1992.
3. G. Kaiser, *A Friendly Guide to Wavelets*, Birkhäuser, Boston, 1994.
4. Y. Meyer, *Wavelets and Operators*, Cambridge University Press, Cambridge, 1992.
5. Y. Meyer, *Wavelets Algorithms and Applications*, SIAM, Philadelphia, 1993.
6. G. Strang and T. Nguyen, *Wavelets and Filter Banks*, Wellesley-Cambridge Press, Cambridge, MA, 1996.
7. P. P. Vaidyanathan, *Multirate Systems and Filter Banks*, Prentice-Hall, Englewood Cliffs, NJ, 1993.
8. M. Vetterli and J. Kovačević, *Wavelets and Subband Coding*, Prentice-Hall, Englewood Cliffs, NJ, 1995.
9. G. Walter, *Wavelets and Other Orthogonal Systems with Applications*, CRC Press, Boca Raton, FL, 1994.
10. M. V. Wickerhauser, *Adapted Wavelet Analysis from Theory to Software*, AK Peters, Wellesley, MA, 1994.

II. Related Data Compression Books.

1. R. J. Clarke, *Transform Coding of Images*, Academic Press, San Diego, CA, 1985.
2. A. Gersho and R. M. Gray, *Vector Quantization and Signal Compression*, Kluwer Academic Publishers, Boston, 1992.
3. R. M. Gray, *Source Coding Theory*, Kluwer Academic Publishers, Boston, 1990.
4. N. J. Jayant and P. Noll, *Digital Coding of Waveforms*, Prentice-Hall, Englewood Cliffs, NJ, 1984.
5. W. Kou, *Digital Image Compression Algorithms and Standards*, Kluwer Academic Publishers, Boston, 1995.

6. W. P. Pennebaker and J. L. Mitchell, *JPEG Still Image Compression Standard*, Van Nostrand Reinhold, New York, 1993.

III. Related Spline Books.

1. C. de Boor, *A Practical Guide to Splines*, Springer-Verlag, New York, 1978.
2. C. K. Chui, *Multivariate Splines*, CBMS Series 54, SIAM, Philadelphia, 1988.
3. S. Karlin, *Total Positivity*, Stanford University Press, Stanford, CA, 1968.
4. G. Nürnberger, *Approximation by Spline Functions*, Springer-Verlag, New York, 1989.
5. I. J. Schoenberg, *Cardinal Spline Interpolation*, CBMS Series 12, SIAM, Philadelphia, 1973.
6. L. L. Schumaker, *Spline Functions: Basic Theory*, Wiley-Interscience, New York, 1981.

IV. Related Edited Volumes.

1. J. J. Benedetto and M. Frazier, eds., *Wavelets: Mathematics and Applications*, CRC Press, Boca Raton, FL, 1994.
2. C. K. Chui, ed., *Wavelets: A Tutorial in Theory and Applications*, Wavelet Anal. and Its Appl. Series 2, Academic Press, Boston, 1992.
3. C. K. Chui, L. Montefusco, and L. Puccio, eds., *Wavelets Theory, Algorithms, and Applications*, Wavelet Anal. and Its Appl. Series 5, Academic Press, Boston, 1994.
4. J. M. Combes, A. Grossmann, and Ph. Tchamitchian, eds., *Wavelets Time-Frequency Methods and Phase Space*, 2nd Edition, Springer-Verlag, New York, 1990.
5. I. Daubechies, ed., *Different Perspectives on Wavelets*, in Proc. Symp. in Appl. Math. 47, American Mathematical Society, Providence, RI, 1993.
6. E. Foufoula-Georgiou and P. Kumar, eds., *Wavelets in Geophysics*, Wavelet Anal. and Its Appl. Series 4, Academic Press, Boston, 1993.
7. T. H. Kroonwinder, ed., *Wavelets: An Elementary Treatment of Theory and Applications*, World Scientific, Singapore, 1993.
8. Y. Meyer, ed., *Wavelets and Applications*, Masson, Paris, 1992.
9. M. B. Ruskai, G. Beylkin, R. Coifman, I. Daubechies, S. Mallat, Y. Meyer, and L. Raphael, eds., *Wavelets and Their Applications*, Jones and Barlett, Boston, 1992.
10. L. L. Schumaker and G. Webbs, eds., *Recent Advances in Wavelet Analysis*, Wavelet Anal. and Its Appl. Series 3, Academic Press, Boston, 1994.

V. Special Journal Issues.

1. *Adv. in Comput. Math.*, Special Issue on Multiscale Techniques, W. Dahmen, ed., Vol. 1, 1995.
2. *IEEE Trans. Inform. Theory*, Special Issue on Wavelet Transforms and Multiresolution Signal Analysis, I. Daubechies, S. Mallat, and A. S. Willsky, eds., March 1992.

3. *IEEE Trans. Signal Processing*, Special Issue on Wavelets and Signal Processing, P. Duhamel, P. Frandrin, T. Nishitani, A. H. Tewfik, and M. Vetterli, eds., December 1993.

4. IEEE Proceedings, Special Issue on Wavelets, J. Kovačević and I. Daubechies, eds., April 1996.

VI. Related Research Papers.

1. L. Andersson, N. Hall, B. Jawerth, and G. Peters, *Wavelets on closed subsets of the real line*, in Recent Advances in Wavelet Analysis, L. L. Schumaker and G. Webb, eds., Academic Press, Boston, 1994, pp. 1–61.

2. P. Auscher, *Ondelettes fractales el applications*, Thèse de Doctorat, University Paris-Dauphine, 1989.

3. G. Battle, *A block spin construction of ondelettes, Part* I: *Lemarié functions*, Comm. Math. Phys., 110 (1987), pp. 601–615.

4. G. Beylkin, R. R. Coifman, and V. Rokhlin, *Fast wavelet transforms and numerical algorithms* I, Comm. Pure Appl. Math., 43 (1991), pp. 141–183.

5. P. J. Burt and E. H. Adelson, *The Laplacian pyramid as a compact image code*, IEEE Trans. Comm., 31 (1983), pp. 532–540.

6. C. K. Chui and J. M. De Villiers, *Spline-wavelets with arbitrary knots on a bounded interval: Orthogonal decomposition and computational algorithms*, CAT Report 359, Texas A&M University, College Station, TX, 1995.

7. C. K. Chui and H. Diamond, *A natural formulation of quasi-interpolation by multivariate splines*, Proc. Amer. Math. Soc., 99 (1987), pp. 643–646.

8. C. K. Chui and C. Li, *A general framework of multivariate wavelets with duals*, Appl. and Comp. Harmonic Anal. (ACHA), 1 (1994), pp. 368–390.

9. C. K. Chui and E. Quak, *Wavelets on a bounded interval*, in Numerical Methods of Approximation Theory, Vol. 6, D. Braess and L. L. Schumaker, eds., Birkhäuser-Verlag, Basel, 1992, pp. 53–75.

10. C. K. Chui and X. L. Shi, *On multi-frequency wavelet decompositions*, in Recent Advances in Wavelet Analysis, L. L. Schumaker and G. Webb, eds., Academic Press, Boston, 1994, pp. 155–189.

11. C. K. Chui and J. Z. Wang, *On compactly supported spline-wavelets and a duality principle*, Trans. Amer. Math. Soc., 330 (1992), pp. 903–915.

12. C. K. Chui and J. Z. Wang, *High-order orthonormal scaling functions and wavelets give poor time-frequency localization*, Fourier Anal. and Appl., 2 (1996), pp. 415–426.

13. C. K. Chui and J. Z. Wang, *A study of compactly supported scaling functions and wavelets*, in Wavelets, Images, and Surface Fitting, P.-J. Laurent, A. Le Méhauté, and L. L. Schumaker, eds., AK Peters, Wellesley, MA, 1994, pp. 121–140.

14. A. Cohen, I. Daubechies, and J. C. Feauveau, *Biorthogonal bases of compactly supported wavelets*, Comm. Pure Appl. Math., 45 (1992), pp. 485–500.

15. A. Cohen, I. Daubechies, and P. Vial, *Wavelets on the interval and fast wavelet transforms*, Appl. and Comp. Harmonic Anal. (ACHA), 1 (1993), pp. 54–81.

16. R. Coifman, Y. Meyer, S. Quake, and M. V. Wickerhauser, *Signal processing and compression with wavelet packets*, in Progress in Wavelet Analysis and Applications, Y. Meyer and S. Roques, eds., Observatoire Midi-Phyrénées de l'Université Paul Sabatier, Editions Frontieres, 1992, pp. 77–93.

17. I. Daubechies, *Orthonormal bases of compactly supported wavelets*, Comm. Pure Appl. Math., 41 (1988), pp. 909–996.

18. Y. F. Dehery, M. Lever, and P. Urcum, *A MUSICAM source codec for digital audio broadcasting and storage*, in Proc. IEEE Int. Conf. ASSP, Toronto, May 1991, pp. 3605–3608.

19. T. Ebrahimi, *Perceptually derived localized linear operators: Application to image sequence compression*, These 1028, Ecole Polytechnique Fedeale de Lausanne, 1992.

20. D. Gabor, *Theory of communication*, J. IEE. London, 93 (1946), pp. 429–457.

21. J. Goswami, A. Chan, and C. K. Chui, *On solving first-kind integral equations using wavelets on a bounded interval*, IEEE Antennas and Propagation, 43 (1995), pp. 614–622.

22. A. Grossmann and J. Morlet, *Decomposition of Hardy functions into square integrable wavelets of constant shape*, SIAM J. Math. Anal., 15 (1984), pp. 723–736.

23. A. Haar, *Zur theorie der orthogonalen funktionensysteme*, Math. Ann., 69 (1910), pp. 331–371.

24. N. J. Jayant, J. Johnston, and B. Safranek, *Signal compression based on models of human perception*, Proc. IEEE, 81 (1993), pp. 1385–1422.

25. M.-J. Lai, *On Strömberg's spline-wavelets*, Appl. and Comp. Harmonic Anal. (ACHA), 1 (1994), pp. 188–193.

26. D. LeGall, *MPEG: A video compression standard for multimedia applications*, Comm. ACM, 34 (1991), pp. 46–58.

27. P. G. Lemarié, *Ondelettes à localisation exponentielles*, J. Math. Pures Appl., 67 (1988), pp. 227–236.

28. T. Lyche and K. Mørken, *Spline-wavelets of minimum support*, in Numerical Methods of Approximation Theory, Vol. 6, D. Braess and L. L. Schumaker, eds., Birkhäuser-Verlag, Basel, 1992, pp. 177–194.

29. S. Mallat, *Multiresolution representation and wavelets*, Ph.D. thesis, University of Pennsylvania, Philadelphia, 1988.

30. Y. Meyer, *Ondelettes et fonctions splines*, Séminaire EDP, École Polytechnique, Paris, December 1986.

31. Y. Meyer, *Principe d'incertitude, bases Hilbertiennes et algèbres d'opérateurs*, Séminaire Bourbaki 662, 1985–1986.

32. E. Quak and N. Weyrich, *Decomposition and reconstruction algorithms for spline wavelets on a bounded interval*, Appl. and Comp. Harmonic Anal. (ACHA), 1 (1994), pp. 217–231.

33. A. Said and W. Pearlman, *A new fast and efficient image codec based on set partitioning in hierarchical trees*, IEEE Trans. Circuits and Systems for Video Tech., 6 (1996), pp. 243–249.

34. H. C. Schweinler and E. P. Wigner, *Orthogonalization methods*, J. Math. Phys., 11 (1970), pp. 1693–1694.

35. J. M. Shapiro, *Embedded image coding using zerotrees of wavelet coefficients*, IEEE Trans. Signal Proc., 41 (1993), pp. 3445–3462.

36. M. J. Shensa, *The discrete wavelet transform: Wedding the à Trous and Mallat algorithms*, IEEE Trans. Signal Proc., 40 (1992), pp. 2464–2482.

37. Y. Shoham and A. Gersho, *Efficient bit allocation for an arbitrary set of quantizers*, IEEE Trans. ASSP, 36 (1988), pp. 1445–1453.

38. J. O. Strömberg, *A modified Franklin system and higher order spline systems on $\mathbb{R}^n$ as unconditional bases for Hardy spaces*, in Proc. Conference in Honor of Antoni Zygmund, Vol. II, W. Beckner, A. P. Calderón, R. Fefferman, and P. W. Jones, eds., Wadsworth, NY, 1981, pp. 475–493.

39. M. Unser, A. Aldroubi, and M. Eden, *On the asymptotic convergence of B-spline wavelets to Gabor functions*, IEEE Trans. Inform. Theory, 38 (1992), pp. 864–871.

40. P. P. Vaidyanathan, *Quadrature mirror filter banks, M-band extensions and perfect reconstruction techniques*, IEEE ASSP Mag., 4 (1987), pp. 4–20.

41. M. Vetterli and C. Herley, *Wavelets and filter banks: Theory and design*, IEEE Trans. ASSP, 40 (1992), pp. 2207–2232.

42. J. L. Walsh, *A closed set of normal orthogonal functions*, Amer. J. Math., 45 (1923), pp. 5–24.

43. G. Zweig, *Basilar membrane motion*, in Cold Spring Harbor Symposia on Quantitative Biology, Cold Spring Harbor Laboratory, 1976, pp. 619–633.

Due to the vast amount of literature on the subject of wavelets and their applications to signal analysis, it would take a great effort even to list all of the major publications. Only those research articles that are quoted in the Summary and Notes section are listed here. Fortunately, the lists of references in the wavelet books, edited volumes, and special journal issues (under I, IV, and V above) are fairly extensive.

Subject Index